Membrane and Membrane Reactors Operations in Chemical Engineering

Membrane and Membrane Reactors Operations in Chemical Engineering

Special Issue Editor

Adolfo Iulianelli

MDPI • Basel • Beijing • Wuhan • Barcelona • Belgrade

MDPI

Special Issue Editor
Adolfo Iulianelli
University of Calabria
Italy

Editorial Office
MDPI
St. Alban-Anlage 66
4052 Basel, Switzerland

This is a reprint of articles from the Special Issue published online in the open access journal *ChemEngineering* (ISSN 2305-7084) from 2018 to 2019 (available at: https://www.mdpi.com/journal/ChemEngineering/special_issues/Membrane_Reactors_Operations).

For citation purposes, cite each article independently as indicated on the article page online and as indicated below:

LastName, A.A.; LastName, B.B.; LastName, C.C. Article Title. *Journal Name* **Year**, *Article Number*, Page Range.

ISBN 978-3-03921-022-0 (Pbk)
ISBN 978-3-03921-023-7 (PDF)

Contents

About the Special Issue Editor

Adolfo Iulianelli (Ph.D.) holds a degree in Chemical Engineering (2002) from the University of Calabria (Italy), where he also obtained his PhD in Chemical and Materials Engineering in 2006. Nowadays, he is a Researcher at the Institute on Membrane Technology of the Italian National Research Council (CNR-ITM). He is author or co-author of more than 170 scientific contributions in the form of international peer-reviewed articles in ISI journals (h-index = 29, Scopus database), patents, proceedings of national and international conferences, as well as more than 30 peer-reviewed book chapters. Furthermore, he is Reviewer of more than 30 international ISI journals and has been the Invited and Keynote Speaker of several international conferences and training schools, amongst others. Furthermore, A.I. has been a member of the Organizing and Scientific Committee of several international conferences.

A.I. is the Editor of various scientific and international books about hydrogen, membrane technology, and renewable sources, and is Associate Editor of the *International Journal of Membrane Science & Technology*, Sectional Editor of *Journal of New Developments in Chemistry—Green Chemistry*, Sectional Editor of *Current Alternative Energy*, member of the editorial board of several scientific journals, and serving as Guest Editor for *International Journal of Hydrogen Energy*, *Membranes*, and *ChemEngineering*.

A.I is qualified as Associate Professor with expertise in 1) Chemical Industrial Process and Plants, 2) Systems, Methodologies & Technologies of Chemical and Process Engineering, and 3) Fundamentals of Chemistry Technologies.

His research interests include membrane reactors, fuel cells, gas separation, hydrogen production from reforming reactions of renewable sources exploitation, and green chemistry.

Preface to "Membrane and Membrane Reactors Operations in Chemical Engineering"

In the last four decades, membrane technology has largely contributed to the valorization of process intensification strategy in several strategic engineering sectors, demonstrating the high potentialities of membrane operations as an alternative approach to conventional processes.

In the field of chemical engineering, high process efficiency and easy operation, high product selectivity and permeability, elevated compatibility in integrated membrane systems, energy saving and environmentally friendly processes, and membrane and membrane reactor operations represent a well-established scientific and industrial reality, as also reported in the various works collected in this Special Issue.

In this Special Issue, Rahimpour and co-authors (Water and Wastewater Treatment Systems by Novel Integrated Membrane Distillation (MD)) have reviewed the recent state-of-the-art and sophisticated advances in membrane distillation technology for wastewater treatment. Di Profio and co-authors (Ionic Liquid Hydrogel Composite Membranes (IL-HCMs)) presented a novel experimental route for the preparation of hydrogel composite membranes for utilization as membrane contactors in desalination applications.

Dittmeyer and co-authors (Experimental Investigation of the Gas/Liquid Phase Separation Using a Membrane-Based Micro Contactor) investigated the gas/liquid phase separation of CO_2 from a water–methanol solution at the anode side of a micro direct methanol fuel cell using hydrophobic polytetrafluoroethylene as a membrane microcontactor. Cassano and co-authors (A Multivariate Statistical Analyses of Membrane Performance in the Clarification of Citrus Press Liquor) performed statistical analysis on the experimental behaviors of polyvinylidene fluoride membranes applied in the clarification of citrus press liquor.

Marino et al. (Hydrogen and Oxygen Evolution in a Membrane Photoreactor Using Suspended Nanosized Au/TiO_2 and Au/CeO_2) proposed a method for one-step hydrogen and oxygen separation through a photocatalytic membrane reactor using a modified Nafion membrane. Through experiments, Morico et al. (Solar Energy-Assisted Membrane Reactor for Hydrogen Production) studied a pilot-scale membrane reformer coupled with solar-assisted molten salt-heating to generate hydrogen, also proposing an economic analysis of its industrial feasibility at reduced environmental impact.

Caravella et al. (Dry Reforming of Methane in a Pd–Ag Membrane Reactor: Thermodynamic and Experimental Analysis) performed an experimental campaign on the CO_2 reforming of methane in a catalytic Pd-based membrane reactor, including a detailed thermodynamic analysis, demonstrating the benefits of this membrane-integrated reaction process while making the production of syngas more efficient and with additional environmental advantages. To conclude, Holgado and Alique (Preliminary Equipment Design for On-Board Hydrogen Production by Steam Reforming in Palladium Membrane Reactors) presented the design of an on-board hydrogen production Pd-based membrane reactor integrated to a PEM fuel cell, demonstrating the feasibility of a one-step process for vehicle applications.

Last but not least, the Editor of this Special Issue would like to thank all the authors for their excellent work and acknowledge their contribution to the success of this project.

Adolfo Iulianelli
Special Issue Editor

chemengineering

MDPI

Review

Water and Wastewater Treatment Systems by Novel Integrated Membrane Distillation (MD)

Parisa Biniaz, Niloofar Torabi Ardekani, Mohammad Amin Makarem and Mohammad Reza Rahimpour *

Department of Chemical Engineering, Shiraz University, Shiraz 71345, Iran; parisabiniaz@shirazu.ac.ir (P.B.); nano_physic@yahoo.com (N.T.A.); amin.makarem@gmail.com (M.A.M.)
* Correspondence: rahimpor@shirazu.ac.ir; Tel.: +98-917-713-9296

Received: 30 September 2018; Accepted: 7 January 2019; Published: 15 January 2019

Abstract: The scarcity of freshwater has been recognized as one of the main challenges people must overcome in the 21st century. The adoption of an environmentally friendly, cost-effective, and energy-efficient membrane distillation (MD) process can mitigate the pollution caused by industrial and domestic wastes. MD is a thermally driven process based on vapor–liquid equilibrium, in which the separation process takes place throughout a microporous hydrophobic membrane. The present paper offers a comprehensive review of the state-of-the-art MD technology covering the MD applications in wastewater treatment. In addition, the important and sophisticated recent advances in MD technology from the perspectives of membrane characteristics and preparation, membrane configurations, membrane wetting, fouling, and renewable heat sources have been presented and discussed.

Keywords: membrane distillation; wastewater treatment; membrane configuration; fouling renewable heat sources

---

1. Introduction

Freshwater scarcity and the excessive consumption of water have been regarded as serious challenges over past decades. Several contributing factors such as an increasing population, improving living standards, agricultural sector growth, and industrialization have threatened a further reduction in the water level and given rise to this crisis [1]. Based on the type of industry, a vast amount of wastewater containing salinity and organic compounds such as arsenic, fluoride, cadmium, chromium, mercury, manganese, lead, etc., have been produced. Discharging these contaminant elements above their effluent standard has exerted catastrophic effects on aquatic and terrestrial habitats and human health [2]. To address this issue, several treatment technologies have been investigated by scientists such as reverse osmosis, disinfection, granular filtration, gravity separation, coagulation-flocculation, air stripping and aeration ion exchange, adsorption, and membrane filtration [3]. Among all the conventional techniques under study, the membrane process has become highly popular due to the potential benefits associated with the technology.

Generally, in a membrane, some particular substances are selectively allowed to pass through while others are retained (retentate phase) [4]. The permeating compounds pass through the membrane based on a driving force such as a pressure gradient, concentration gradient, temperature gradient, or electrical gradient [5,6]. This phenomenon emerges from membrane module characterization (pore size, pore shape), membrane surface characteristics (porosity, charge/hydrophobicity) and membrane configuration (geometry, dimensions) [7]. To put it another way, membrane separation processes applied in wastewater treatment are categorized as the isothermal and non-isothermal process. The former includes concentration-driven membrane processes (pervaporation and membrane extraction), pressure-driven membrane processes (microfiltration, ultrafiltration, nanofiltration, and reverse osmosis) and electrically driven membrane processes (electrodialysis, electrophoresis) while the latter is a thermally driven membrane process named membrane distillation (MD) [8–10]. According to the literature, among all the membrane processes, MD has been perfectly able to treat water with an extremely high level of salinity [11,12] and hazardous contaminants [13]. In this context, extensive research has been conducted by scientists and researchers all over the world over recent years. Drioli et al. [14] investigated the current and prospective role of membrane engineering in attaining the objectives of a process intensification strategy to improve the efficiency and sustainability of novel membrane processes including MD.

The present study assesses the evolution of membrane distillation in wastewater treatments. The work investigates the characteristics, material, module, and different configurations of MD applied in water treatment as well as covering the fouling and wetting phenomena. Furthermore, the benefits and limitations of MD processes, economic analysis, and future research directions of interest have been pointed out. While various review papers focusing on MD technology have been conducted by researchers, most of them provide a full membrane perspective, without being highly focused on the novel MD membrane designs and process configurations. In this critical review, the authors aim to review the recent advances in MD technology in terms of low-grade or renewable heat sources, such as waste heat from industrial processes which reduce transmembrane heat loss and increase the proportion of heat recovered from the permeate stream. Moreover, nontraditional anti-fouling processes and recently developed membranes prepared from surface modifications of polymers and nanomaterials such as plasma surface modification and electrospinning are investigated thoroughly.

2. Membrane Distillation

2.1. History of MD Process

Membrane distillation was patented in 1963 [15] and Findley published the first MD paper in 1967 in the "Industrial & Engineering Chemistry Process Design Development" journal [16]. Nonetheless, the MD process did not attract considerable interest until the early 1980s when more efficient module membranes such as Gore-Tex became accessible [17]. The term MD originates from the significant similarity between the MD method and typical distillation process since both of them operate based on the liquid/vapor equilibrium. Moreover, in both processes feed stream is heated by the energy source to obtain the necessary potential heat of vaporization [18]. Being a practical and effective wastewater treatment, MD has been the topic of worldwide investigation by many researchers and scientists. Besides, the interest in MD processes has rapidly grown over the recent years. Table 1 gives up-to-date and interesting information concerning the application of MD in wastewater treatment based on several patents published from 2016 to 2018.

Table 1. List of published patents in the application of membrane distillation (MD) in wastewater treatment (2016 to 2018).

Patent	Highlights
Desulfurization waste water zero-discharging treatment technology for coal-fired power plants Publication number: CN105712557A Publication date: 2016-06-29	The present invention relates to a zero-discharging treatment device for desulfurization wastewater that consists of a nanofiltration system, calcium removal sedimentation pool set, a heavy metal and magnesium removal pool set, an evaporating crystallizer, and a membrane distillation system. Crystal salt gained from evaporating crystallization can be entirely recycled, and treatment and operating cost are extremely reduced.
A membrane distillation system that is used for concentration of desulfurization waste water Publication number: CN204981458U Publication date: 2016-01-20	The utility model reveals the MD system that is applied for the handling of the concentration of desulfurized wastewater, with this considered to have some benefits such as a simple structure, safe operation, and a smaller area.
Salt-containing wastewater treatment system Publication number: CN205133326U Publication date: 2016-04-06	The present utility model includes a salt-containing wastewater treatment system that contains an electrodialysis device for receiving and handling salted wastewater, which is connected to the MD system.
Process and system for produced water treatment Publication number: US20170096356A1 Publication date: 2017-04-06	This invention relates to a system and process for produced water treatment. A heat exchanger, a vacuum tank, an adsorption-desorption, and MD crystallization process are in this process combination. The efficiency related to costs of maintenance and energy consumption is considered.
Modularly installed energy-saving membrane distillation wastewater treatment device and method Publication number: CN106865663A Publication date: 2017-06-20	The present invention represents a modularly installed energy-saving MD wastewater treatment device which consists of a post-treatment and primary treatment module. The proposed device assembles easily and has suitable energy-saving effects for the treatment of different types of sewage.
Energy-saving membrane distillation effluent treatment plant of modularization installation Publication number: CN206635064U Publication date: 2017-11-14	The utility model relates to an energy-saving MD sewage treatment plant of modularization installation and consists of post-processing and primary treatment modules. The applied used equipment in the device is energy conserving, convenient, and effective for all types of effluent treatment.
High -concentration organic wastewater treatment system Publication number: CN206318843U Publication date: 2017-07-11	The present utility model reveals a highly concentrated organic wastewater treatment system that consists of a liquid bath of raw material, anaerobic biological treatment device, and a positive osmotic membrane. Highly concentrated organic wastewater from a bioreactor is handled in the combined MD system. This system proposes greater recovery of the pure water rate and reduces the concentration of organic matters.
Multi-stage submerged membrane distillation water treatment apparatus and a resource recovery method Publication number: US20170313610A1 Publication date: 2017-11-02	This investigation related to a submerging multi-stage membrane distillation water treatment device.

3

Table 1. *Cont.*

Wastewater treatment system Publication number: CN207243660U Publication date: 2018-04-17	This utility model shows a wastewater treatment system that consists of an MD unit connected to magnetism loaded flocculation unit. This proposed wastewater treatment system has some benefits like simple equipment, efficient sewage treatment, and a lower energy consumption.
Concentrated decrement device of vacuum membrane distillation wastewater that frequently flows backwards Publication number: CN207734625U Publication date: 2018-08-17	This invention concerns a zero-discharging wastewater treatment device, which is particularly related to a vacuum membrane distillation (VDM). MD has some effective benefits, for instance, it enhances the service life, and shows great abilities to treat wastewater with high salinity.
Slot-type solar sea water desalination device based on membrane distillation Publication number: CN107720863A Publication date: 2018-02-23	The invention reveals a slot-type solar seawater desalination device integrated with membrane distillation. The required heat energy of the process is supplied from solar energy emitted by a slot-type condenser mirror, in which solar energy is reflected and condensed onto an arc heat collection tube.
Porous membrane for membrane distillation, and method for operating membrane distillation module Publication number: WO2018174279A1 Publication date: 2018-09-27	A membrane distillation device, with a hydrophobic porous hollow fiber membrane, and a condenser for condensing water vapor is invented for water treatment. The membrane has an average pore diameter of 0.01–1 μm.
Hollow fiber membrane module for direct contact membrane distillation-based desalinization Publication number: WO2018195534A1 Publication date: 2018-10-25	The invention is a desalination system by direct contact membrane distillation integrated with a cylindrical cross-flow module comprising high-flux composite hydrophobic hollow fiber membranes. A model is developed and directed to the system and shows the observed water vapor production rates for various feed brine temperatures at different feed brine flow rates.
A membrane distillation technique and method for treating radioactive waste water systems Publication number: CN108597636A Publication date: 2018-09-28	The invention shows a seed film distillation procedure and technology for processing radioactive waste including (pretreatment, preheating, membrane separation, condensation process) by accumulating the wastewater
Multistage immersion type membrane distillation water treatment apparatus and a resource recovery method using the same number of oil resources Publication number: KR101870350B1 Publication date: 2018-06-22	This invention provides a multistage immersion-type membrane distillation water treatment system and a viable resource recovery technique applying the same number of oil resources which can substantially reduce the heat energy.

2.2. Definition of MD

MD is a thermally driven treatment process in which the thermal gradient is generated across a microporous hydrophobic membrane [19]. Simultaneously, the process can be operated by low-grade heat and/or waste including solar energy [20], geothermal energy [21], wind, tidal, and nuclear energy, or low-temperature industrial streams [22]. It should be noted that the process is driven by the vapor pressure difference between the permeable hydrophobic membrane pores. In other words, volatile vapor molecules are allowed to pass through the MD while non-volatile compounds are retained on the retentate stream. The permeated volatile vapors are then collected or condensed by various techniques. Finally, completely pure products that are theoretically 100% free from solid, harmful substances and non-volatile contaminants are produced [23]. Figure 1 illustrates a schematic diagram of the MD process.

As the graph reveals, volatile vapor molecules in the hot feed which are vaporized at the liquid/vapor interface are able to pass through the pores of the membrane. The liquid feed, on the other hand, is prevented from transporting through the membrane pores. This phenomenon is mainly because of the hydrophobic nature of the MD membrane and its surface tension. Therefore, it is important to note that the dry pores must not be wetted by the liquid feed which is directly in contact with the hydrophobic membrane [24–26].

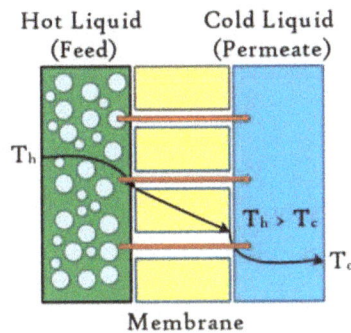

Figure 1. Schematic diagram of the MD process [10].

2.3. Limitation and Benefits of MD Process

In comparison with other conventional membrane separation systems, MD separation process brings tremendous benefits such as having low operating temperatures, being cost-effective by applying waste heat and renewable energy sources, being able to treat wastewater with a high level of purity, and being less likely to suffer from membrane fouling [18,27,28]. Therefore, these remarkable features make MD an attractive method for wastewater treatment, seawater desalination, and so many other industrial applications including environmental purification, in the food industry [29], in medicine [30,31], and in the production of acids, etc. However, employing MD in industry is limited by some significant challenges like the risk of total or partial membrane wetting [32] or not being commercially available on a large scale [33]. Regarding previous studies, the choosing of suitable membranes and the energy efficiency are the two main factors that must be taken into account when applying the MD process [18,23]. Table 2 presents a detailed overview of the positive and negative aspects of the MD process.

Table 2. A comprehensive overview of the positive and negative points of the MD process.

Advantages	Disadvantages	Reference
Low operating temperature (the process liquid is not essentially heated up to the boiling temperatures)	Lower permeate flux compared to other commercialized separation processes, such as RO.	[9,19,33–35]
Lower hydrostatic pressure required compare to pressure-driven membrane separation processes such as reverse osmosis (RO).		[9,19,33]
High rejection (99–100%) for macromolecules, non-volatile compounds (colloids, salts), and inorganic ions. In fact, 100% separation happens, theoretically.	High susceptibility of permeate flux to temperature and concentration polarization effects, partial or total pore wetting, and membrane scaling and fouling.	[19,36,37]
Lower requirements on the mechanical properties of the membrane.		[18,33]
Larger pore size and less chemical interaction between process solution and membrane lead to less fouling.	High heat loss (by conduction) and energy consumption	[33,36]
Alternative low-grade energy sources like waste heat, solar energy, and geometrical heat can be utilized.	Pore wetting risk	[8,38]
The possibility to combine with some other separation processes in order to build an integrated separating system, like an RO unit or ultrafiltration	Unclear economic and energy costs for different MD applications and configurations, just when waste heat is available MD becomes cost competitive.	[8,33]
An efficient method to eliminate heavy metals and organic from wastewater.		[33]
It is an effective and safe process to remove radioactive waste.	The lack of commercially available MD modulus manufactured for large-scale applications and high-performance membrane.	[33,39]
MD is able to work with a saturated solution or high solute concentration in a liquid stream		[33]
Fewer vapor spaces needed in comparison with common distillation process so MD can be used at a smaller scale.	Having less producers of MD technology	[8,33]
Reduced sensibility to concentration polarization.		[8,33,38]
High concentration polarization or osmotic pressure does not limit performance.		
Having low cost and less sophisticated installation and construction (because of lower operating temperature and pressure), leads to a full level of automation.	Limitations of MD permeate flux, due to a further mass transfer resistance caused by trapped air through the membrane	[18,33]
Being less sensitive to membrane pollution or concentration polarization and without a pretreatment stage.		[8,33,36]

2.4. Membrane Characteristics

Hydrophobicity is the fundamental necessity for an MD membranes process. Therefore, the membranes must be fabricated with original or modified hydrophobic polymers with low surface energies. Moreover, the membrane applied in the MD should have a low resistance to mass transfer and low thermal conductivity to prevent heat loss across the membrane. In addition, the membrane should have good thermal stability in high temperatures, and good chemical resistance to acids and bases. High permeability is another significant feature that a membrane should possess in order to be applied in the MD process. To satisfy this feature, the membrane surface layer must be as thin as possible so that the vapors are allowed to pass through the membrane in a short period of time. Another notable characteristic of a membrane is a high liquid entry pressure (LEP). LEP is regarded as the minimum hydrostatic pressure in an MD system that prevents the liquid solutions from penetrating into the membrane pores. A sufficiently high LEP can be achieved by applying a membrane material with high hydrophobicity and a small maximum pore. In addition, the surface porosity and pore size of the membrane must be as large as possible [18,40].

2.5. Membrane Materials and Modules

The most popular micro-porous hydrophobic membranes are commercially fabricated with polypropylene (PP), polyvinylidene fluoride (PVDF), polytetrafluoroethylene (PTFE), polyethylene (PE), inorganic materials, and carbon nanotubes (CNTs) available in plate and frame, hollow fiber, tubular, spiral wound, and flat sheet modules. Currently, PTFE with unique features such as high hydrophobicity and strong resistance against severe operating conditions has dominated the commercial and laboratory applications of MD [23,41,42]. Figure 2 makes a comparison between different membrane modules with their positive and negative points [9]. Table 3 provides information about the characteristics of commercial membranes commonly applied in the MD process.

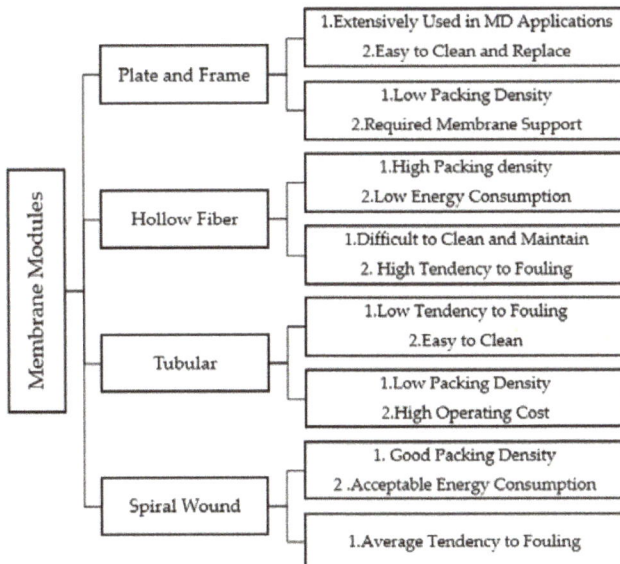

Figure 2. Advantage and disadvantages of membrane modules.

Table 3. Commercial membranes applied in MD (membrane thickness, δ; porosity, ε; liquid entry pressure of water, LEPw).

Membrane Trade Name	Material	Manufacturer	δ (μm)	ε (%)	LEP$_W$ (kPa)	Reference
TF200	PTFE */PP **	Gelman	178	80	282	
TF450	PTFE/PP	Gelman	178	80	138	[8,18,43]
TF1000	PTFE/PP	Gelman	178	80	48	
PT20	PTFE/PP	Gore	64 ± 5	90 ± 1	3.68 ± 0.01	[8]
PT45	PTFE/PP	Gore	77 ± 8	89 ± 4	2.88 ± 0.01	
TS1.0	PTFE/PP	Osmonics Corp.	175	70	-	
TS22	PTFE/PP	Osmonics Corp.	175	70	-	[18]
TS45	PTFE/PP	Osmonics Corp.	175	70	-	
Taflen	PTFE/PP	Gelman	60	50	-	
FGLP	PTFE/PE	Millipore	130	70	280	
FHLP	PTFE/PE ***	Millipore	175	85	124	
GVHP	PVDF ****	Millipore	110	75	204	
PV22	PVDF	Millipore	126 ± 7	62 ± 2	2.29 ± 0.03	[8,44]
PV45	PVDF	Millipore	116 ± 9	66 ± 2	1.10 ± 0.04	
HVHP (Durapore)	PVDF	Millipore	140	75	105	
GVSP	PVDF	Millipore	108	80	-	[18]
Gore	PTFE	Gore	64	90	368	
Gore	PTFE	Gore	77	89	288	
Teknokrama	PTFE	Teknokrama	-	80	-	
Teknokrama	PTFE	Teknokrama	-	80	-	
Teknokrama	PTFE	Teknokrama	-	80	-	
G-4.0-6-7	PTFE	GoreTex Sep GmbH	100	80	463	
Sartorious	PTFE	Sartorious	70	70	-	
MD080CO2N	PP	Enka Microdyn	650	70	-	
MD020TP2N	PP	Enka Microdyn	1550	70	-	[8,18]
Accurel®	PP	Enka A.G.	400	74	-	
Celgard X-20	PP	Hoechst Celanese Co	25	35	-	
Accurel® S6/2	PP	AkzoNobel	450	70	1.4	
Enka	PP	Sartorious	100	75	-	
Enka	PP	Sartorious	140	75	-	[18]
3MA	PP	3M Corporation	91	66	-	
3MB	PP	3M Corporation	81	76	-	
3MC	PP	3M Corporation	76	79	-	
3MD	PP	3M Corporation	86	80	-	
3ME	PP	3M Corporation	79	85	-	
Membrana	PP	Membrana, Germany	91	-	-	
PP22	PP	Osmonics Corp.	150	70	-	
Metricel	PP	Gelman	90	55	-	
Celgard 2400	PP	Hoechst Celanese Co.	25	38	-	
Celgard 2500	PP	Hoechst Celanese Co.	28	45	-	
EHF270FA-16	PE	Mitsubishi	55	70	-	

* Polytetrafluoroethylene; ** Polypropylene; *** polyethylene; **** Polyvinylidene fluoride.

3. Conventional MD Configurations

A MD process can be categorized into four basic configurations, which plays a fundamental role in separation efficiency and processing cost. Figure 3 shows a schematic diagram of various conventional configurations including direct contact membrane distillation (DCMD), air gap membrane distillation (AGMD), sweeping gas membrane distillation (SGMD), and vacuum membrane distillation (VMD) [40,45–47].

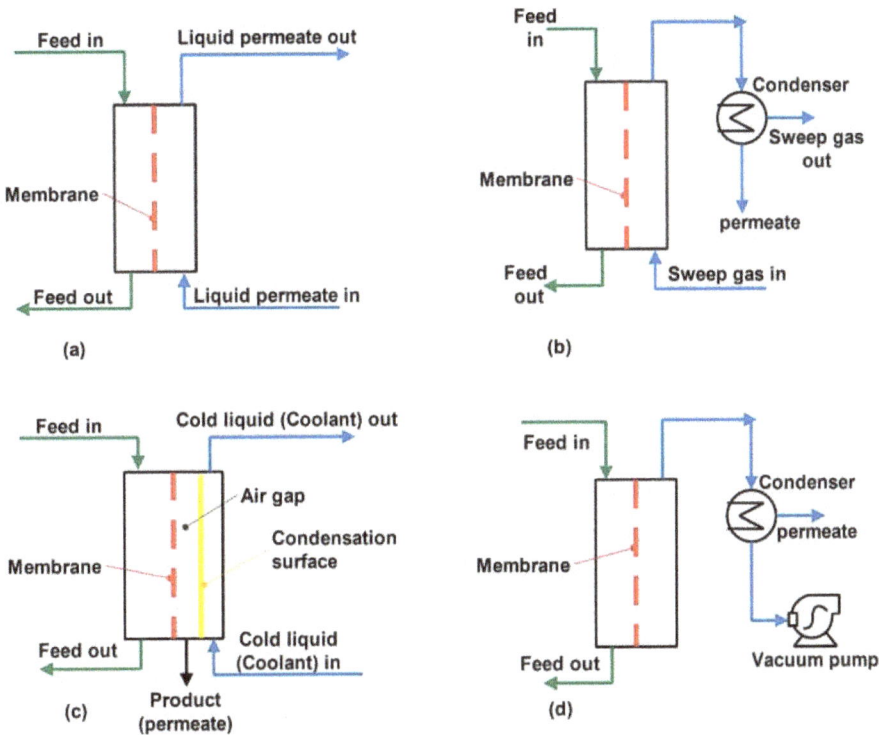

Figure 3. Schematic diagram of conventional membrane configurations (**a**) direct contact membrane distillation (DCMD); (**b**) sweeping gas membrane distillation (SGMD); (**c**) air gap membrane distillation (AGMD); (**d**) vacuum membrane distillation (VMD).

3.1. Direct Contact Membrane Distillation (DCMD)

DCMD is the simplest MD configuration, in which a liquid phase (feed) with high temperature is in direct contact with the hot side of the membrane surface, and a cold aqueous phase is in direct contact with the permeate side. Therefore, volatile compounds evaporate at the hot liquid/vapor interface at the feed side. Having been passed through the membrane pores, the vapor phase will be condensed in the cold liquid/vapor interface at the permeate side. It is notable that the vapor pressure difference is induced by the temperature difference across the membrane and the hydrophobic nature of the membrane prevents the feed from penetrating through the membrane. Despite its simplicity, the conduction heat loss associated with this process is higher than in other configurations. Membrane modules in DCMD could be shell-and-tube or plate-and-frame employed under cross-flow or longitudinal flow [38,48–52].

3.2. Air Gap Membrane Distillation (AGMD)

In this configuration, the evaporator channel is similar to that in DCMD. However, an air gap, which is the controlling factor for the heat and mass transfers, is interposed among the membrane and the cooled surface. The evaporated volatile compounds pass through the membranes and the air gap and then condense onto the cold surface. A significant point about this configuration is that the condensation surface separates the permeate (distillate) from the cold liquid (coolant). Therefore, the cold liquid can be other liquid like cold feed water. The AGMD configuration has the highest energy efficiency among the other configurations and the applied membrane could be both flat sheet and hollow fiber. Furthermore, the used condensation surface is usually a thin dense polymer or metal film [53–59]. Nonetheless, when it comes to popularity, the AGMD process lags behind DCMD and VMD processes because of its complicated module design [60]. Kalla et al. [60] comprehensively reviewed the most recent developments in the AGMD process. Based on their investigation, material gap membrane distillation, double stage AGMD unit, conductive gap membrane distillation, superhydrophobic condenser surface, multi-stage and multi-effect membrane distillation, modified air gap membrane distillation, tangent and rotational turbulent inlet flow, and vacuum assisted air gap membrane distillation process are recent advancements in AGMD process. Woldemariam et al. [61] presented an exergetic analyzing (exergy evaluations are necessary tools for analyzing the performance of separation systems, including those featuring MD.) of AGMD systems at a laboratory and pilot scale. The energy efficiency results indicated that the materials of the condensation plate play a crucial role in optimizing the performance of MD systems such as heat transfer across modules. Stainless steel and polypropylene were considered as the appropriate materials in this regard.

3.3. Sweeping Gas Membrane Distillation (SGMD)

In SGMD, which is also known as air stripping membrane distillation, an inert gas (stripping gas) is applied at the permeate side of the membrane as a carrier to sweep the vapor or collect vapor molecules from the membrane surface. Similar to AGMD, a gas barrier decreases the heat loss and significantly increases mass transfer making SGMD a process with promising future perspectives [62,63]. Nonetheless, SGMD generates a small volume of permeate vapors while needing large volumes of sweep gas and external condensers, consequently incurring extra expenses. Therefore, the process has received little attention in comparison with other MD configurations such as DCMD [19,64]. Applying metallic hollow fibers or a coating membrane with polydimethylsiloxane could significantly enhance the water vapor permeate flux up to 40% in sweep gas membrane distillation [65]. Moore et al. [66] developed a non-steady process model to simulate an SGMD system integrated with solar thermal and photovoltaic power for the desalination of drinking water. The economic analysis indicated that the optimized proposed technology for cost recovery over a 20-year service life is 84.7 $/m^3, which is more than alternative sources water costs. Therefore, future work on SGMD is necessary to make this system economically competitive.

3.4. Vacuum Membrane Distillation (VMD)

In the VMD configuration, a vacuum is created by a pump at the permeate side of the membrane module. Then, an external condenser is used as for AGMD if the permeate stream is the product. In addition, the vapor pressure difference is formed by continuous removal of the vapor permeate from the vacuum chamber. To form the driving force, the created vacuum must be less than the saturation pressure of volatile compounds in the aqueous feed. For the VMD configuration, the conduction heat loss is negligible and membrane wetting is avoidable [9,18,67–69]. Table 4 makes a comparison between different conventional MD configurations and represents the merits and demerits of each conventional process.

Table 4. The merits and demerits of conventional MD configurations.

Method of Treatment	Advantages	Disadvantages	Reference
DCMD	Simplest operation Least required equipment Simplest MD configuration	Not suitable for removing non-volatile organics and dissolved gasses (water must be permeating flux) Highest heat loss by conduction among other configurations	[33,38,48]
AGMD	High flexibility in MD configuration Less conductive heat loss Less tendency to fouling High flux Without wetting on the permeate side	Creation of additional resistance to mass transfer Hard module designing Minimum obtained output ratio	[8,9,55]

Table 4. *Cont.*

SGMD	A suitable configuration for removing contaminant (volatile component and dissolved gasses) Without wetting from the permeate side Lower thermal polarization	Large condenser needed due to the small volume of permeate diffuses in a large sweep gas volume Low flux	[8,9,33,70,71]
VMD	Negligible conductive heat loss High flux Suitable for aroma compounds recovery	Pore wetting risk Higher fouling Vacuum pump and external condenser	[8,9,33]

4. New MD Configurations

Several novel configurations with low energy consumption and improved permeation flux have been developed by scientists and researchers. A brief review of newly proposed MD configurations is now presented.

4.1. Thermostatic Sweeping Gas Membrane Distillation (TSGMD)

The AGMD and SGMD processes can be combined in a process named thermostatic sweeping gas membrane distillation (TSGMD). As Figure 4 clearly illustrates, the inert gas is passed through the gap between the condensation surface and the membrane. Part of the vapor is condensed on the condensation surface (AGMD) and the remainder is condensed over the external condenser (SGMD) [18,72]. This phenomenon basically takes place to minimize the temperature of the sweeping gas, which increases significantly along the membrane module length. In other words, the presence of the condensation surface in the permeate side decreases the temperature of the sweeping gas, which leads to an enhancement in the driving force and the wastewater treatment performance [33,47]. Condensate production in the TSGMD can be increased by enhancing the membrane area, recycling cool air back to the membrane module, and decreasing the airflow across the cooling fins.

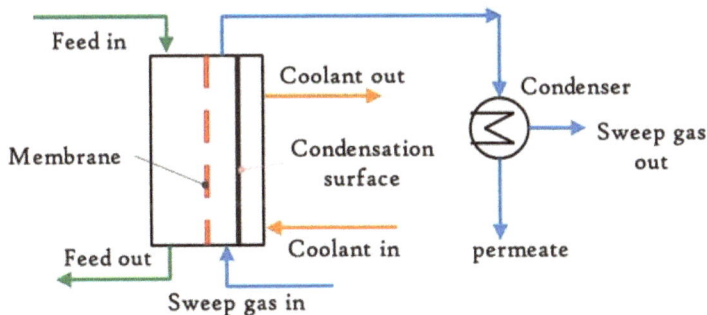

Figure 4. Schematic diagram of thermostatic sweeping gas membrane distillation (TSGMD) [73].

Tan et al. [74] developed a novel SGMD system coupled with a thermoelectric heat pump (TSGMD) to improve the energy efficiency of the water treatment system. The results indicated that applying a T-SGMD system is capable to double the condensate production per unit energy consumed. Furthermore, condensate production in the proposed system can be increased by enhancing the membrane area, recycling cool air back to the membrane module, and decreasing the airflow across the cooling fins. Cool air recycle could affect the condensate flux without a serious loss of cooling in comparison with other tested factors during the operation of the T-SGMD. More importantly, the TSGMD system was able to increase the condensate produced per unit energy without a significant loss in the cooling capacity per unit energy input. This process can be simulated by using a multicomponent Stefan–Maxwell mathematical model. Based on the model analysis, Rivier et al. [62] concluded that since the selectivity of TSGMD is far from unity and can be manipulated by changing operational conditions, this process is suitable for separating azeotropes. Furthermore, the thermal conductivity of the sweeping gas in TSGMD is four-fold lower than that of the membrane, and a small mass transfer resistance exists in the cold chamber due to the gaseous bulk. In one study, the formic acid-water azeotropic mixture was separated by this module [75]. Both experimental and modeling results suggested that TSGMD can considerably shift selectivity with respect to vapor-liquid equilibrium (VLE) data, and the system can be successfully used for the separation processes.

4.2. Multi-Effect Membrane Distillation (MEMD)

This configuration is an AGMD module with an internal heat recovery system based on the concepts of multi-effect, which is suitable for seawater desalination. The cold feed is placed under the condensation surface as a coolant to condense the permeated vapor compounds as well as to absorb heat. The pre-heated feed solution is heated once again before entering the feed channel [19,76]. Figure 5 shows a schematic diagram of the MEMD process. The source of heating for this system should be between 50 to 100 °C and the utilized membrane is usually micro-porous PTFE. It is reported that the module has very low specific energy consumption (between 56 to 100 kWh/m^3) [77].

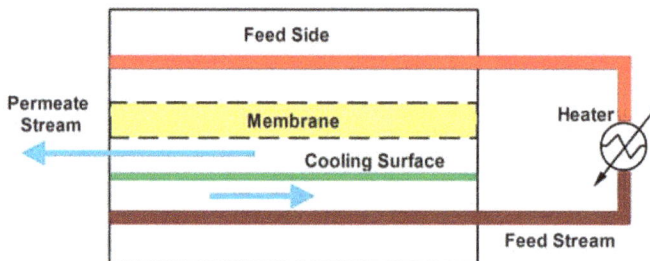

Figure 5. Schematic diagram of the multi-effect membrane distillation (MEMD) [19].

4.3. Vacuum Multi-Effect Membrane Distillation (V-MEMD)

VMEMD shares a similar concept with the multi-effect membrane distillation (MEMD) except for the vacuum enhancement. The system includes a heater, multiple evaporation–condensation stages, and an external condenser. Consequently, the distillate is created in both condensation surfaces and inside the external condenser. In addition, the vacuum condition is developed at the air gap between membrane and condensation surfaces to eliminate the excess air/vapor from the process. In this context, memsys is a state of the art technology and relatively new configuration based on vacuum multi-effect membrane distillation. This highly efficient technology consists of a novel internal heat recycling concept that results in a significant reduction in energy consumption [19,69,78–80]. Figure 6 shows a schematic diagram of the V-MEMD system. The performance of this type of membrane is affected by changing heating conditions, cooling, and feed. The main factors for its optimization and scale-up are the number of stages and the size of each stage. In an experimental study, Zhao et al. [80] found that heating and cooling temperatures are the most important factors affecting the module flux and efficiency. By comparing this module with common technologies, one can conclude that it can provide better heat and mass transfer rates [81]. Furthermore, Mohamed et al. [82] reported that flow rates are also important factors affecting the module performance. They suggested that the system performance can be improved more by performing the following tips: separating distillate and brine tanks, utilizing magnetic or float flow meters, using a separate pump for heating recirculation, and improving the heat isolation or recovering the heat loss in the brine stream.

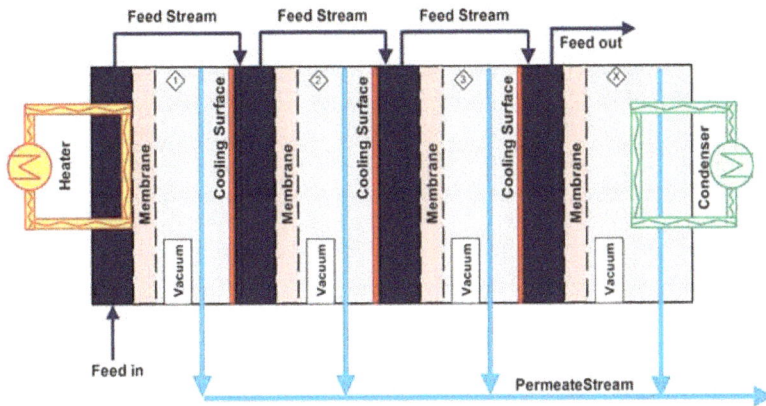

Figure 6. Schematic diagram of vacuum multi-effect membrane distillation (V-MEMD) [19].

4.4. Material-Gap Membrane Distillation (MGMD)

Since AGMD has shown the lowest permeation flux among all configurations, a new and efficient configuration named material gap membrane distillation (MGMD) was developed and designed by researchers to compensate for the weakness of AGMD. In this membrane module, the air gap is filled with either nonconductive materials such as porous support, sand, and sponge (polyurethane) or conductive materials such as the metal mesh. Therefore, the vapor compound flux increases up to 200–800% [83–86]. Figure 7 shows a schematic diagram of the MGMD system.

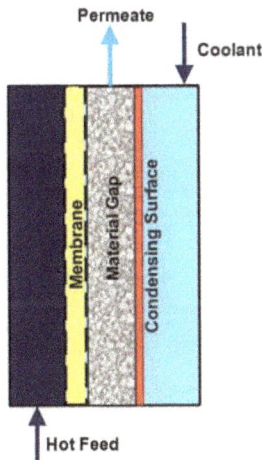

Figure 7. Schematic diagram of material gap membrane distillation (MGMD) [83].

To simulate large-scale module conditions, it is essential to perform the experiments at low-temperature gradients across the membrane. This is because at this condition, the heat recovery is maximized and the water vapor flux would be at a minimum [81].

4.5. Permeate-Gap Membrane Distillation (PGMD)

The combination of DCMD configuration and AGMD module is regarded as permeate-gap membrane distillation (PGMD) or liquid-gap membrane distillation (LGMD). In this configuration, the additional compartment between the membrane and the condensation surface is filled with a static cold liquid solution or with permeate. It is also notable that applying PGMD leads to a higher-surface-related permeate output in comparison with AGMD. This is mainly because, in AGMD, the diffusion resistance of the air layer acts as an obstacle in the process. However, PGMD has greater heat loss than AGMD [83]. Figure 8 shows a schematic diagram of the PGMD system. Winter et al. [85] proposed PGMD with internal heat recovery that was achieved by separating the distillate and coolant. As a result, any liquids such as the feed water can be utilized as the coolant. Therefore, one can place PGMD between AGMD and DCMD to lower the sensible heat transfer to the permeate, but at the cost of greater heat loss. By accepting the superior performance of PGMD over AGMD, Swaminathan et al. [84] reported that the countercurrent flow of the pure water in the gap to the cold stream results in the highest energy efficiency, and increasing the gap conductivity enhances the permeate production.

Figure 8. Schematic diagram of permeate-gap membrane distillation (PGMD) [86].

5. Application of MD

Generally, MD has been applied in various areas including desalination, the chemical industry, the food industry, the textile industry, pharmaceutical, and biomedical industries and the nuclear industry. Table 5 provides more information about each application.

Table 5. The application of the MD process in different industries.

Area	Application	MD Configuration	Reference
Chemical industry	Removing volatile organic compounds from water Acid concentrating Crystallization Azeotropic mixtures separation	VMD DCMD SGMD AGMD	[33,87–89]
Desalination	Producing pure water from brackish water	VMD DCMD SGMD AGMD	[33,34,90,91]
Food industry (Juice and Dairy)	Juice concentrating Processing of milk Temperature sensitive materials	VMD DCMD AGMD	[33,92–95]

Table 5. *Cont.*

Textile industry	Dye removal Wastewater treatment	VMD DCMD	[33,96–98]
Pulp and paper industry	Removing sodium sulfate, organic and inorganic compounds, adsorbable organic halogens (AOX), color, phenolic compounds, and chemical oxygen demand (COD) from wastewater	DCMD	[99–101]
Pharmaceutical and biomedical industries	Wastewater treatment Water removing from protein and blood solutions	DCMD	[30,33,102,103]
Nuclear industry	Producing pure water Wastewater treatment Radioactive solutions concentrating	DCMD VMD	[31,33,104,105]
Gold mining	Reusing mining effluents Removing hazardous metals and ions such as sulfate from mining effluents	DCMD	[106]
Bioethanol production plants	Recovery of ethanol from scrubber-water	AGMD	[107]

The Application of MD in Water and Wastewater Treatment

MD technology has been extensively used in the purification of wastewater produced from various industries, in order to recover valuable compounds or make wastewater less dangerous to the environment. However, in comparison with other membrane processes, MD is more difficult to apply on an industrial scale because of some serious economic and engineering problems.

When it comes to desalination, many various types of technologies are available including thermal- and membrane-based desalination processes. The thermal-based group comprises processes such as multistage flash distillation (MSF); multi-effect distillation (MED); and, single- or multiple-effect evaporation (SEE/MEE) systems, which can be coupled to mechanical or thermal vapor compression (MVC/TVC). Membrane technologies include reverse osmosis (RO), forward osmosis (FO), electrodialysis (ED), and membrane distillation (MD). Onishi et al. comprehensively reviewed the main advantages and disadvantages of each process [108].

RO has been considered as the most economical and the least energy intensive technology for large-scale seawater desalination, followed by MED and MSF [23,109,110]. However, the unique characteristics of MD have made this process an excellent option with high efficiency. For instance, MD, in particular, DCMD, has enormous potential in the desalination of highly saline wastewaters where MD fluxes can remain comparatively high, much higher than those for RO. Moreover, in small-scale applications where the quality of water is not suitable for currently established technologies such as RO-based processes, MD is an effective alternative. This process could also be co-located with industrial facilities and power generation systems to take advantage of the waste heat and low-cost thermal energy to produce high-quality water. In addition, MD is a potential treatment candidate for combining with other separation techniques such as RO, ED, crystallization, and bioreactors to enhance water recovery and decrease the amount of concentrate requiring disposal. Therefore, MD has a practical application in water treatment with zero (or near zero) liquid discharge and can be more economical than other established thermal processes in zero-liquid discharge applications. The permeate with extremely high quality in the MD process compared to RO permeate can also offer considerable benefits particularly when purified water is required as boiler feed [19,23,111]. In mining industries, the process combining ultrafiltration (UF) and reverse osmosis (RO) is widely applied for wastewater treatment, in which 80% of COD, more than 95% chroma and almost all the ferrous irons and bacteria can reject significantly. Nonetheless, the brine discharge and the water recovery ratio (limited to around 30% to 60%) of the RO process remain serious issues in this regard. Therefore, MD could be proposed to address these problems by enhancing the water recovery ratio and recovering minerals [8,112,113].

Lokare et al. [114] investigated the synergies and potential of DCMD for wastewater treatment produced during gas extraction from unconventional (shale) reservoirs in Pennsylvania (PA).

An exhaust stream from Natural Gas Compressor Station (NG CS) was used as the waste heat source for DCMD operation providing a feasible option to treat high salinity generated water. They developed an ASPEN Plus simulation of DCMD using fundamental heat and mass transfer equations and the literature correlations to optimize the design and operation of large-scale saline water desalination and estimated the energy requirements of the process. The minimum temperature gradient of 10 °C between the permeate-side and feed stream was used to achieve the optimum membrane area when several membrane modules are provided in series. According to obtained results, the amount of available waste heat of NG CS regardless of the produced water salinity is much higher than the amount of waste heat needed to concentrate produced water in PA to 30% salinity. Moreover, the results indicated that DCMD is able to concentrate all the produced water in PA utilizing NG CS waste heat. Nonetheless, the economic probability of the process must be assessed to determine major cost drivers and barriers. Ali et al. [115] evaluated the integration of a microfiltration and membrane distillation process for water treatment and minerals recovery from produced water. The results indicated that the integrated process offers the opportunity of converting generated water into salt and freshwater highly efficiently and also minimizes the issue of waste disposal. Nonetheless, to make the system commercially available, better arrangements for separating crystals of various salts should be made. Boukhriss et al. [116] simulated and experimentally studied an AGMD membrane distillation pilot for the desalination of brackish water and seawater with zero liquid discharged. The theoretical model was generated using Matlab and verified utilizing pilot-scale experimental data. Their investigation showed that the AGMD configuration is capable of producing desalinated water with zero liquid discharged even at a low hot-fluid supply temperature of 25 °C, which makes the system feasible to be coupled with low-temperature heat sources such as a solar collector. Baghbanzadeh et al. [117] investigated a zero thermal energy input membrane distillation (ZTIMD) process which was also a zero-waste system. The required thermal driving force for the process was provided by using the warm seawater of the surface as the feed and the cold water at the bottom of the sea as the coolant. The innovative concept of their invention revolutionized the seawater desalination industry. This is mainly because ZTIMD was revealed to be economically more efficient than current seawater desalination processes by simulations based on a single-pass DCMD system. In other words, under the optimum conditions, the proposed ZTIMD process could provide pure water with a cost of \$0.28/m^3 at the particular energy consumption of 0.45 kW h/m^3, which is notably lower than that of the main current seawater desalination methods such as RO (\$0.45–2.00/m^3). A novel water desalination method which couples thermal membrane distillation (TMD) with reverse osmosis (RO) was developed by Huang et al. [118]. They proposed a water–energy integration process with the strong nexus of water and energy. Furthermore, a dual-objective model was formulated to analyze the system thermodynamically and optimize the process under the objective function of minimizing fuel and freshwater consumption. The sensitivity analysis of the heat-to-power demand ratio revealed that the RO-TMD coupling water desalination process is superior to traditional single RO at a high heat-to-power demand ratio in terms of minimizing freshwater and fuel consumption. In the proposed novel water–energy integration system, the fuel and freshwater consumptions were reduced by 1.7% and 21.0%, respectively, compared with those of the conventional system.

A comprehensive summary of more MD applications in wastewater treatment on a laboratory-scale is presented in Table 6. Although the number of research papers published in the MD application in wastewater treatment has increased significantly over the past few years, tremendous effort should be taken to design and fabricate novel membrane modules to permit a successful industrial application of this separation technique. It is worth mentioning that, various excellent membrane modules have become available in the market recently. Therefore, in the near future, a MD process with industrial applications may become increasingly available.

Table 6. Summary of MD applications in wastewater treatment.

Feed	Membrane Configuration	Membrane Material	Contaminant	Removal Efficiency (%)	Scale	Reference
Radioactive wastewater (SrCl2)	VMD	PP	Sr^{2+}	Over 99.60	laboratory-scale	[104]
Metal solution (salts of Co (II), Zn (II), Cu (II), Ni (II), Cd (II) and Pb (II))	VMD	Poly(vinylidenefluoride)-titanium tetraisopropoxide PVDF-TTIP	Heavy metals	Total removal	laboratory-scale	[119]
Distilled water and crude oil	VMD	PVDF	Total Organic Carbon (TOC)	93.4-97	laboratory-scale	[120]
Olive Mill WasteWater (OMWW)	DCMD & VMD	PP	Polyphenols TOC	99.6 89 and 99.6	laboratory-scale	[121]
Industrial textile wastewater	DCMD	PVDF-Cloisite 15A nanocomposite	Colour Total Dissolved Solids(TDS) Chemical Oxygen Demand(COD)	95.3 93.7 90.8	laboratory-scale	[122]
Synthetic dye solution	DCMD	PVDF modified by ethylene glycol (EG)	RB5	99.86	laboratory-scale	[98]
Highly saline radioactive wastewater	DCMD	PP	Nuclides (Co(II), Sr(II), Cs(I)) and boron (B)	>99.97%	laboratory-scale	[123]
Synthetic wastewater and Seawater	Osmotic membrane bioreactor (OMBR)—(DCMD) hybrid system	PTFE active layer and a PP supporting layer	30 trace organic contaminants	>90%	laboratory-scale	[37]
Geothermal water	AGMD	PP, PTFE and PVDF	Boron	99.5%	laboratory-scale	[21]
High salinity hydraulic fracturing produced water (HFPW)	Combined Electrocoagulation (EC) and DCMD	Ethylene chlorotrifluoroethylene (ECTFE)	Turbidity, Total suspended solids (TSS) and TOC	96%, 91% and 61%, respectively	laboratory-scale	[124]
Industrial dyeing wastewater	DCMD combined with physicochemical and biological treatment	PTFE and PVDF	COD and color removal	96% and 100% respectively	laboratory-scale	[49]
Saline oily wastewater	DCMD	PVDF modified with silica nanoparticles and Polystyrene (PS) microspheres	Oil and gas emulsified wastewater	Highly desirable	laboratory-scale	[25]
Mining wastewater	VMD	PVDF membrane was coated by Hyflon AD materials (Hyflon AD40L, Hyflon AD40H)	Mining waste	Highly efficient	laboratory-scale	[126]
Bentazon herbicide solutions	VMD	PTFE	Bentazon	Very effective	laboratory-scale	[127]
Fermentation wastewater	DCMD	PP	COD, TOC	95%	laboratory-scale	[128]

6. Fouling Phenomenon in MD Process

The term "fouling" has been regarded as a complex phenomenon which is an inevitable part of each membrane process and adversely affects membrane performance. Generally, fouling is the precipitation and accumulation of various foulants such as particles and dissolved components on the external surface or inside the membrane as pore blocking. Figure 9 clearly illustrates external surface fouling and internal fouling pore blocking.

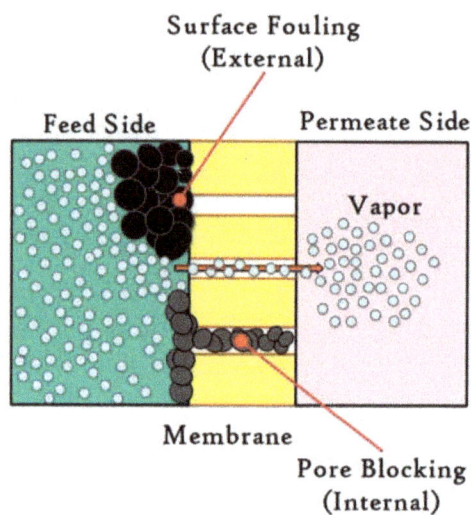

Figure 9. Schematic diagram of surface fouling (external) and pore blocking (internal) [35].

A fouling layer puts on extra thermal and hydraulic resistance to the process and decreases the temperature difference across the membrane, which means a sharp reduction in the driving force. Consequently, the permeate flux decreases drastically. If fouling does not address this properly, it will contribute to membrane damage, early membrane replacement or even shutdown of the operation [35]. The main fouling phenomena commonly occur during water and wastewater treatment and are categorized based on the foulant type as organic, inorganic (scaling), biological, and particle fouling. When suspended solids and metal hydroxide in source water accumulate on the membrane surface and inside the membrane pores, forming a cake layer, the phenomenon is known as particle fouling. Scaling is regarded as the precipitation of inorganic salts presented in source water such as calcium carbonate, calcium sulfate, silicate, NaCl, calcium phosphate, $BaSO_4$, $SrSO_4$, ferric oxide, iron oxide, aluminum oxide, inside the membrane pores, which leads to bulk, pore plugging/clogging, and membrane crystallization. Furthermore, the adsorption of natural organic matter (NOM) such as HA, fulvic acid, protein, polysaccharides, and polyacrylic polymers on the membrane has been considered as organic fouling which contributes to gel formation of the macromolecular compounds and membrane wetting. As well as that, when various aquatic organisms such as fungi, sludge, algae, yeast, and micro-organisms in source water form a biofilm on the membrane, the fouling is named biofouling [35,129]. Nevertheless, mostly, a combination of several types of fouling mechanisms occurs in actual MD processing, as opposed to a single fouling mechanism which makes the problem more complicated to address. It is worth mentioning that membrane fouling is expected to be less devastating in MD, due to the absence of hydraulic pressure in such processes compared to pressure-driven membrane processes. Nonetheless, the continuing presence of the membrane in the highly concentrated feed solution to meet the pure liquid discharge makes the MD process vulnerable to membrane fouling.

Bush et al. [130] compared the performance and fouling behavior of MD and nanofiltration (NF) processes applying silica-saturated water from 225 mg/L to 600 mg/L SiO_2 to illustrate the potential differences in the silica scaling behavior and its impacts on MD performance compared to a pressure-driven membrane process. The results showed that salt rejection during MD was >99.8% for all solutions tested and was unaffected by scaling, while rejection during NF was between 78–90% and tended to decrease after scaling. NaOH solution at pH > 11 was used to clean the fouled membranes for both processes, which was extremely effective at the restoring water flux but unable to remove the silica scale layer completely. Tow et al. [131] analyzed the fouling and scaling behavior of RO, forward osmosis (FO), and DCMD using a single membrane module under the same hydrodynamic conditions (flux and cross-flow velocity). During fouling experiments, calcium sulfate was used as a model inorganic foulant and alginate was utilized as a model organic foulant. Based on their results, FO showed the greatest scaling resistance while MD tolerated organic fouling much better than FO and RO. Although FO and MD each indicated a higher resistance to one type of foulant, neither process outperformed RO in the resistance to complex fouling including organic and inorganic fouling.

Typically, the characteristics of foulants (concentration, molecular size, solubility, diffusivity, hydrophobicity, charge), water (solution chemistry, pH, ionic strength, presence of organic/inorganic matters), and membrane (hydrophobicity, surface roughness, pore size, surface charge, surface functional groups), as well as operational conditions (flux, solution temperature, flow velocity), can significantly affect the fouling formation phenomenon [61,132]. In this regard, various approaches are employed by researchers in order to detect and prevent membrane fouling summarized in Table 7. Shan et al. developed a versatile approach for designing an amphiphobic membrane surface [133]. During this method, a biomimetic system was investigated to design an amphiphobic surface with a unique structure and controllable wettability. A commercial PVDF was modified via superhydrophobic nanocoating using polydopamine (PDA) followed by the fluorination of 1H,1H,2H,2H-perfluorodecanethiol. The proposed amphiphobic membrane indicated excellent superhydrophobicity with a water contact angle of 167.6° + 0.27° as well as remarkable chemical and thermal stability under severe conditions. Another striking feature about this membrane was its outstanding anti-fouling capability with higher flux and great salt rejection in the long-term DCMD process, which exhibits promising potentials for industrial applications.

Table 7. Summary of detection and prevention methods for membrane fouling.

Detection Method(s)	Prevention Process(es)	Reference
• Permeate flux decline • Scanning electron microscopy (SEM) • Energy dispersive X-ray spectroscopy (EDS) to evaluate the elemental composition of a fouled layer • X-ray powder diffraction (XRD) to analyze and evaluate the crystalline nature of inorganic, organic, polymers, metals, or composite materials • Atomic force microscopy (AFM) is applied to characterize the surface of the membrane • SEM–EDS/TEM (transmission electron microscopy)–EDS • Atomic absorption spectroscopy (AAS) analysis to determine the fouling composition using the absorption of optical radiation (i.e., light) by free atoms in the gaseous state • Contact angle (CA) (the angle between the liquid drop and the horizontal surface, an angle below 90° shows hydrophilic behaviour, while an angle above 90° shows hydrophobicity) • Membrane autopsy • Fourier transform infrared (FTIR) to obtain the infrared (IR) spectra of the sample for evaluating and identifying the chemical bonds and molecular structure of organic molecules and analyzing of organic and inorganic functional groups on the membrane surface • Attenuated total reflectance-fourier transform infrared spectroscopy (ATR-FTIR) to analyze organic and inorganic fouling • Gel permeation chromatography or gel filtration chromatography (HPSEC) or using flow field-flow fractionation FlFFF to determine the size or molecular weight distribution of organic matters • Liquid-chromatogram organic carbon detection (LC-OCD) • Zeta potential (to determine the electrokinetic phenomena of membranes and evaluate the possible interaction between the foulants and membrane surface)	• Membrane flushing (membrane cleaning by regular deionized (DI) water) • Ultrasonic irradiation technique • Chemical cleaning of membrane by acids, alkalis, metal chelating agents, surfactants, enzymes, and oxidizing agents (2 h water, with 0.1 M NaOH or 2–5 wt % HCl solutions or 0.029 M Na_2EDTA and 0.058 M NaOH or citric acid followed by NaOH or by HCl) • Pretreatment (coagulation, multimedia filtration, sonication, deep-bed filtration, pH changes boiling/thermal water softening, chlorination, degasification,) • Gas bubbling • Increasing the feed flow rate • Temperature and flow reversal • Using anti-fouling membranes (such as membrane surface modification by applying different superhydrophobic coatings such as sodium alginate hydrogel, TiO_2 nanoparticles, a mixture of polydimethylsiloxane (PDMS) and hydrophobic SiO_2 nanoparticles) • Using antiscalants (chemical additives) such as condensed polyphosphates, organophosphonates, and polyelectrolytes	[35,61,128,129,132,134–150]

Table 7. *Cont.*

- Tensile strength
- Direct visualization
- EXSOD (ex-situ scale observation detector) for real-time crystal monitoring of membranes
- Autopsy
- Optical laser sensor method for investigation of the deposit thickness on a membrane
- Ultrasonic time–domain reflectometry (UTDR) to observe and evaluate the deposition of combined organic and colloidal fouling on the membrane surface as well as providing physical characteristics of the media where the waves travel and also providing the real-time measurement of the location of an interface
- Inductively coupled plasma mass spectrometry (ICP-MS) to determine the concentration (ppm, ppb) of metal and non-metal elements in a fouling layer
- Streak-plate method to determine the number of microorganisms
- A confocal laser scanning microscopy (CLSM) technique is applied to obtain high-resolution optical images with depth selectivity and also to visualize and quantify biofilms in-situ in combination with a fluorescent probe.
- Infrared thermography (IRT) technique is applied to measure the surface temperature and its distribution figure out whether a foulant is metallic or non-metallic in nature
- Excitation Emission-Matrix Fluorescence Spectroscopy (EEM) to detect proteins, acid and other organic materials present in fouled membranes

7. Wetting Phenomenon in MD Process

In the wetting phenomenon, water enters the pores of the membrane and fills them by breaking the surface tension presenting between liquid and vapor on the surface of the membrane [151]. As previously described, mass transfer through the membrane pores only takes place in the vapor phase, from a hot feed solution. Therefore, the liquid feed must not be allowed to penetrate partially or entirely through the dry membrane pores. As Figure 10 clearly illustrates, the degree of membrane wettability may vary according to the area wetted by the liquid. The water might be present only in the external surface layers of membrane pores, or a fraction of pores inside the membrane with the largest diameters might be wetted, or all the pores inside the membrane might be filled by water. To address this issue, various approaches have been developed [152–156]. Extra liquid entry pressure and membrane fouling are two principal causes of membrane wetting. Kim et al. [157] fundamentally investigated the MD process integrated with crystallization in order to prevent inorganic scaling induced by multivalent ions such as barium and calcium in shale-gas-produced water treatment. By utilizing the proposed system, scalant loading was decreased significantly and membrane wetting was mitigated dramatically. Therefore, the total recovery of the process was increased up to 62.5%. Furthermore, experimental results indicated that the pretreatment process for oil and grease before MD application is mandatory for enhancing the stability of water production and sustaining the integrity of permeate water quality.

A comprehensive review of wetting mechanisms, wetting causes, and wetting detection methods was carried out by the authors of [156]. Jacob et al. [26] developed a novel method to visualize and evaluate wetting in the MD system based on the detection of dissolved tracer intrusion, which is capable of detecting all pore wetting mechanisms at different locations on a membrane. The function of the developed method was based on the ex-situ detection of a tracer (salt) intrusion by SEM-EDX, after the performance of the VMD applying a saline solution. Guillen-Burrieza et al. [158] investigated the effects of the MD operation variables on membrane wetting. They concluded that the wetting time and rate are significantly reduced by the parameters enhancing flux. Damtie et al. [159] suggested a new methodology to treat highly polluted industrial wastewater and analyzed the wetting tendency of different membranes. They studied the performance of the most popular commercially available hydrophobic membranes including polytetrafluoroethylene (PTFE), polypropylene (PP), and polyvinylidene fluoride (PVDF). The investigation confirmed that the type of membrane material and membrane pore size greatly influence the process efficiency. Moreover, the relationship between the membrane surface wettability and MD performance is affected by the composition of feed water during the water treatment [160].

Eykens et al. [161] explored an alternative for the traditional hydrophobic membrane materials through the deposition of a hydrophobic coating using vacuum plasma on a commercial hydrophilic membrane with a microporous structure to prevent membrane wetting at high salinity. The required hydrophobicity (>100°) was obtained and membrane wetting was prevented effectively. Chen et al. [162] developed a ZnO nanostructure on alumina hollow fiber membranes with a uniform pore size of 197 nm and a thin wall of 200 μm to enhance the wetting resistance during the DCMD process with a low surface tension feed. The contact angle of the omniphobic hollow fiber (HF) membrane for a 90% v/v ethanol/water mixture was as high as 138.1°. The SEM, EDX, and AFM analyses showed that the omniphobic alumina hollow fiber membranes not only showed extraordinary wetting resistance for desalinating low surface tension wastewaters but also showed a great potential for industrial applications because of the simplicity of scaling-up.

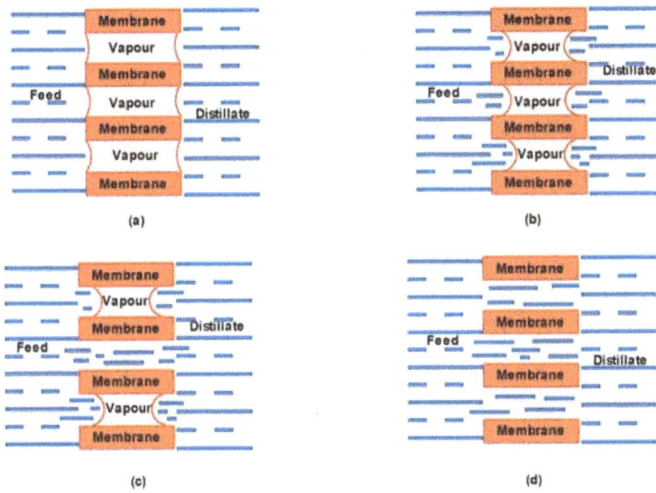

Figure 10. Schematic diagram of membrane wettability stages: (**a**) non-wetted, (**b**) surface wetted, (**c**) partial wetted, and (**d**) fully wetted [35].

8. Novel Approaches to MD Technology

Polymeric porous membranes are traditionally fabricated by conventional methods, most of which contribute to relatively low porosity. Recently, various novel membrane production techniques that have a high porosity above 80% and interconnected open pore structures with a high surface roughness are applied to enhance the membrane performance and provide high flux in MD [163]. Figure 11 illustrates a detailed classification of traditional and novel membrane production methods. Eykens et al. [163] offered a comprehensive definition of each technique and provided advantages and disadvantages of each method in their review paper.

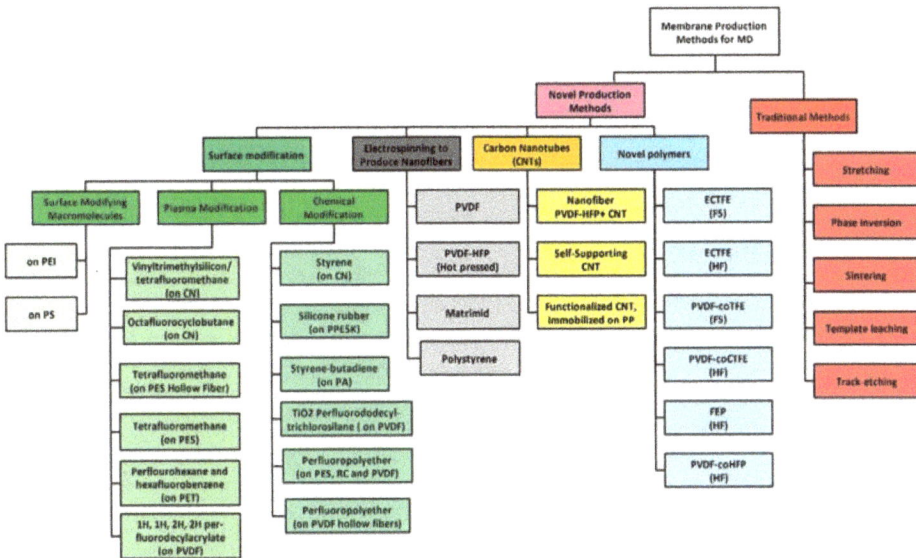

Figure 11. Schematic diagram of various methods for membrane production applied in MD technology [163].

Woo et al. [164] investigated the development and performance of an omniphobic PVDF membrane which was successfully fabricated by electrospinning and modified by tetrafluoromethane (CF$_4$) plasma treatment for water brine treatment with an AGMD system. They studied the effects of various durations of plasma treatment on the characteristics of the nanofiber membrane. The optimum obtained results (treatment duration: 15 min; liquid entry pressure: 187 kPa; flux: 15.28 L/(m^2·h); salt rejection ~100%) demonstrated that the formation of new CF$_2$-CF$_2$ and CF$_3$ bonds after plasma treatment without considerably altering the morphology and physical properties could resist the wetting phenomenon by reducing membrane surface energy and providing omniphobic property for low surface tension liquids such as methanol, mineral oil, and ethylene glycol. Therefore, the proposed omniphobic membrane has great potential to treat water containing high salinity and organic contaminants. An et al. [165] fabricate an amphiphobic PVDF-co-HFP electrospun nanofibrous membrane with excellent anti-wetting properties for the MD process. They applied 1H,1H,2H,2H-perfluorodecyltriethoxysilane (FAS) to fluorinate PVDF-co-HFP fibers followed by a crosslinking process to form a network upon dealcoholization under thermal treatment. Based on their results, the FAS-coated PVDF-co-HFP nanofibrous membranes show excellent stable amphiphobicity with high contact angles of 127° against water and oil even on challenging and critical conditions such as long-term operation, presenting sodium dodecyl sulfate surfactant in the saline feed, or boiling water and strong base and acid etchings. Boo and Elimelech [166] provided a self-heating membrane for MD via CNT Joule heating which enhances the desalination efficiency of high-salinity brines. This novel technology increases the thermal driving force by increasing the temperature of the saline feed stream without the need for external heat. Joule heating that is also known as ohmic heating or resistive heating is the process in which thermal energy is produced by the resistance of a conductor to electron flow. In general, in self-heating membranes, a thin conductive composite layer is formed via a sequential spray coating of CNTs and Polyvinyl alcohol (PVA) on a hydrophobic porous substrate (polytetrafluoroethylene).

More importantly, applying novel renewable energy-driven systems in water treatment has dramatically increased with the aim of energy conservation. Figure 12 illustrates using different renewable energy sources including solar energy, waste heat, and geothermal energy as well as applying a membrane distillation crystallization (MDC) method via the precipitation of crystal salts under supersaturation conditions in a crystallizer. A detailed explanation of each method could be found in the following references [64,167]. Furthermore, MD systems with water recycling and heat regeneration could significantly enhance water recovery and thermal efficiency and, consequently, are capable to meet the actual demand [168]. Long et al. [169] investigated a DCMD system integrated with low-temperature waste heat for water treatment. They developed a modified model characterizing the heat and mass transfer in the DCMD, which was validated by excellent agreement with the experimental data. Based on their study, gain output ration (GOR) and mass recovery rate are two major factors to evaluate the performance of a DCMD system with heat recovery. Lee et al. studied the effects of two different types of seawater-coolant feed (backward feed (BF) and parallel feed (PF)) arrangements in a waste-heat-driven multistage vacuum membrane distillation with regard to the improvement of system performance [170]. Based on their investigations, the proposed system with the BF arrangement is more efficient and economical for freshwater production than the PF arrangement at a smaller number of module stages in terms of the specific thermal energy consumption. Furthermore, they comprehensively analyzed different BF arrangement scenarios and found the optimal number to be a 24-stage VMD desalination system in terms of energy efficiency and cost.

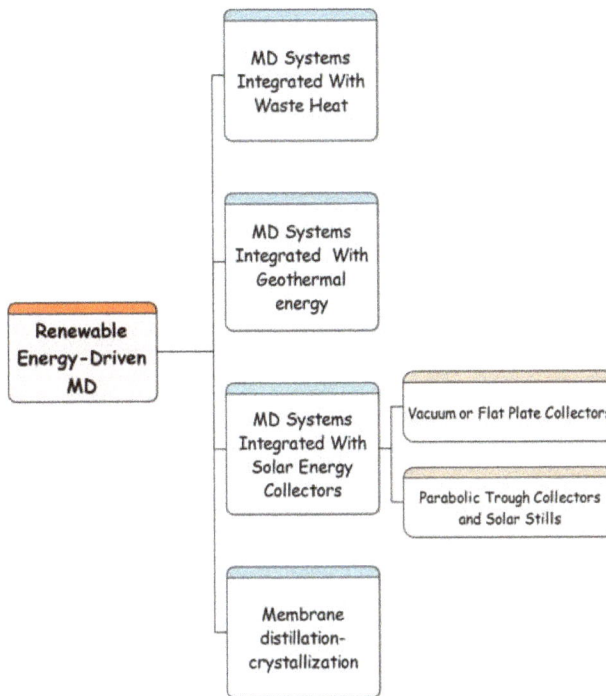

Figure 12. Schematic diagram of various renewable energy-driven MD systems.

9. Economic Analysis of MD Process

Energy-efficient water treatment and desalination processes play a crucial role in enhancing freshwater supplies without imposing considerable strain on scarce resources. By applying low-grade or waste heat, membrane distillation (MD) has shown great potential to augment sustainable water. Nonetheless, economic analyses are essential for the viability of the MD process since a huge amount of energy is used for the water evaporation in order to separate water from non-volatile contaminants and pumping [33]. Furthermore, it should be noted that an energy-efficient MD system effectively applies the thermal gradient for vapor transfer in comparison with conductive heat loss, which is measured by the membrane thermal efficiency of the membrane. In addition, the latent heat of condensation is productively reused in an economical MD system [90]. Recently, the authors of [171] critically examined the crucial factors affecting the energy efficiency of MD processes and explained how future membrane design and process development could considerably boost MD efficiency. They demonstrated that the size of the system can significantly influence the performance of the process. Moreover, they found that enhancing the porosity of the membrane and optimizing its thickness can dramatically increase the MD efficiency. In addition, the configuration of the process plays a leading role in maximizing the latent heat recovery. More importantly, the novelty of membrane materials and surface modification are essential for increasing membrane robustness. Swaminathan et al. [172] comprehensively analyzed a single-stage MD system in terms of the energy efficiency (indicate as a gained output ratio or GOR) and vapor flux for the desalination of feed streams up to a high feed salinity. The system was designed to determine the thickness of the membrane with an optimal cost as well as the size of the system as a function of the ratio of specific costs of heat energy and module area. Based on the obtained results, in the small systems with low salinity, GOR increases and flux decreases with a rise in the membrane area. Furthermore, for the solutions with high salinity, it is essential to determine a critical system size beyond which GOR begins to decrease.

Based on the performance of the MD, the total capital cost of the system, optimum flow conditions, and MD configurations, and the cost of the MD system production may vary from 0.26 to 130 $/m³. Furthermore, the total energy consumption of the process could change from 1 to 9000 kWh/m³ based on the type and size of the system, operating conditions, sources of the provided energy, recovery approaches, and the estimated cost of the procedures. Additionally, by applying waste heat, the production cost of a 30,000 m³/d (capacity plant) MD desalination plant could be reduced from 2.2 to 0.66 $/m³ [173,174].

A comprehensive cost evaluation of a 111 MWe solar power tower (SPT) plant integrated with the DCMD system was investigated by Soomro and [114,175]. The average freshwater production by the proposed MD system was evaluated up to 40,759 L/day with a cost of $0.392/m³. The authors of [176] evaluated the economic feasibility of MD for wastewater treatment by performing a techno-economic assessment (TEA) for a hypothetical 0.5 million gallons per day (MGD) direct contact MD (DCMD) which concentrates produced water from 10% (100,000 mg/L) TDS to 30% salinity. Sensitivity analysis showed that the TDS level of the feed and the price of thermal energy significantly affect the total cost of treating produced water. Furthermore, they revealed that utilizing a source of waste heat could considerably decrease the total cost from $5.70/$m^3_{feed}$ to $0.74/$m^3_{feed}$. Hitsov et al. [177] demonstrated a graphical user interface tool which could design a comprehensive MD system, comprising all of the supporting equipment and capable to estimate the price of the obtained distillate for different distillation configurations at various production scales and concentration factors. They also investigated several case studies that varied from 2 to 1000 m³ of distillate per day, with a final brine salinity up to 20 wt.% and feed temperature up to 80 °C, and demonstrated the optimal system design for each case. The cost of distillation varied from 25 €/m³ (the smallest scale) to 2.1 €/m³ for the largest scale. Soomro and Kim [178] published an economic evaluation of integrating a 50 MWe parabolic-trough (PT) plant with the DCMD system for freshwater production. The economic analysis illustrated that the proposed system could be a sustainable and economical process producing up to 14.33 m³ of freshwater per day at a price of $0.64/m³.

According to a technical economic study conducted by the authors of [106], the capital expenditure (or capital expenses) (CAPEX) and operational expenditure (or operational expenses) (OPEX) of DCMD applied in gold mining effluent treatment (for a membrane lifespan of 1–5 years) are estimated to be US$ 305,483.85 and 0.13 to 0.27 US$/m³, respectively, while the amounts are US$ 575,490.30 and 2.00 to 2.10 US$/m³ for the NF process. This is mainly because the required energy for NF is almost 40 times greater than that for DCMD, due to the need for cooling the feed in the NF process. Moreover, 98% of the thermal energy consumption in DCMD is reduced by applying the residual heat of the effluent. The CAPEX was measured as explained by Hitsov et al., considering a membrane area of 24 m² per module and the capital cost (*Ccap*) was calculated per cubic meter of effluent (*Ccap*/m³), as explained in great detail by Reis et al. [106]. Woldemariam et al. [107] evaluated the CAPEX and OPEX of an industrial-scale district heat-driven MD process for the recovery of ethanol from scrubber water. The economy of the distillation system was obtained from the case studied plant including production rate, the amount of steam used, and other costs such as capital investment. Results of the techno-economic investigation indicated that MD could be a competitive technology for ethanol recovery when the system is supplied by low-grade heat such as waste heat or district heating network.

10. Future Trends and Conclusions

MD is a thermally driven treatment process, which has been perfectly able to treat water containing an extremely high level of salinity and hazardous contaminants. More importantly, MD has the possibility to integrate with other separation processes and renewable energy sources. Nonetheless, few studies were performed on a large scale and with long-term MD application since several challenges such as high energy consumption, fouling, scaling, and pore wetting have limited its commercial application. Therefore, it is essential to fabricate novel membranes with specific characteristics such as low resistance to mass transfer, low thermal conductivity, high thermal stability, high chemical resistance, or a membrane with surface modifications to improve MD performance and characteristics in order to minimize fouling and wetting phenomena and energy consumption, as well as enhancing the permeate flux quality and quantity. Moreover, the development of MD application in wastewater treatment needs to handle more organic and biological fouling, in combination with inorganic scaling. This makes the fouling study more complicated; therefore, more insight into the mechanisms of mixed fouling should be given careful and special attention in the future.

Generally, MD process can be categorized into four basic configurations including direct contact membrane distillation (DCMD), air gap membrane distillation (AGMD), sweeping gas membrane distillation (SGMD), and vacuum membrane distillation (VMD). Additionally, various novel configurations with low energy consumption and improved permeation flux such as TSGMD, MEMD, V-MEMD, MGMD, and PGMD have been proposed recently. Polypropylene (PP), polyvinylidene fluoride (PVDF), polytetrafluoroethylene (PTFE), polyethylene (PE), inorganic materials, and carbon nanotubes (CNTs) are the most popular micro-porous membranes commercially fabricated in the form of plate and frame, hollow fiber, tubular, spiral wound, and flat sheet. However, different novel techniques such as electrospinning and surface modification have been employed recently to produce a membrane with a high porosity of above 80% and enhance the membrane performance by providing high flux.

Author Contributions: The manuscript was written by P.B. and reviewed by all authors. All authors have given approval to the final version of the manuscript.

Funding: This research received no external funding.

Conflicts of Interest: The authors declare no conflict of interest.

Abbreviations and Symbols

AGMD	Air Gap Membrane Distillation
COD	Chemical Oxygen Demand
CN	Cellulose Nitrate
CNT	Carbon Nanotube
DCMD	Direct Contact Membrane Distillation
ECTFE	Poly (ethene-co-chlorotrifluoroethene)
ED	Electrodialysis
FEP	poly (vinylidene fluoride-co-chlorotrifluoroethylene)
FO	Forward Osmosis
FS	Flat sheet
HF	Hollow fiber
LEPW	Liquid Entry Pressure of Water
LGMD	Liquid-Gap membrane distillation
MD	Membrane Distillation
MEE	Multiple-Effect Evaporation
MED	Multiple-Effect Distillation
MGMD	Material-Gap Membrane Distillation
MSF	Multi-Stage Flash
NF	Nanofiltration
NOM	Natural Organic Matter
PES	Polyethersulfone
PET	Poly(ethylene terephthalate)
PGMD	Permeate-Gap Membrane Distillation
PP	Polypropylene
PS	Polysulfone
PPESK	Poly(phthalazinone ether sulfone ketone)
PTFE	Polytetrafluoroethylene

PVA	Polyvinyl alcohol
PVDF	Polyvinylidene fluoride
PVDF-co-CTFE	Pol (vinylidene fluoride-co-chlorotrifluoroethylene(
PVDF-co-HFP	Poly(vinylidene fluoride-co-hexafluoropropylene)
PVDF-co-TFE	Poly(vinylidene fluoride-cotetrafluoroethylene)
RC	Regenerated cellulose
RO	Reverse Osmosis
SGMD	Sweeping Gas Membrane Distillation
SEE	Single-Effect Evaporation
TDS	Total Dissolved Solids
TOC	Total Organic Carbon
TSGMD	Thermostatic Sweeping Gas Membrane Distillation
TSS	Total suspended solids
VMD	Vacuum Membrane Distillation
V-MEMD	Vacuum Multi-Effect Membrane Distillation
ZTIMD	Zero Thermal Input Membrane Distillation
δ	Thickness
ε	Porosity

References

1. Eckardt, N.A.; Cominelli, E.; Galbiati, M.; Tonelli, C. The future of science: Food and water for life. *Am. Soc. Plant Biol.* **2009**. [CrossRef]
2. Bejaoui, I.; Mnif, A.; Hamrouni, B. Performance of reverse osmosis and nanofiltration in the removal of fluoride from model water and metal packaging industrial effluent. *Sep. Sci. Technol.* **2014**, *49*, 1135–1145. [CrossRef]
3. Crittenden, J.C.; Trussell, R.R.; Hand, D.W.; Howe, K.J.; Tchobanoglous, G. *MWH's Water Treatment: Principles and Design*; John Wiley & Sons: Hoboken, NJ, USA, 2012.
4. Rahimpour, M.R. 10–Membrane reactors for biodiesel production and processing. In *Membrane Reactors for Energy Applications and Basic Chemical Production*; Basile, A., Di Paola, L., Hai, F.l., Piemonte, V., Eds.; Woodhead Publishing: London, UK, 2015; pp. 289–312. [CrossRef]
5. Van der Bruggen, B.; Vandecasteele, C.; Van Gestel, T.; Doyen, W.; Leysen, R. A review of pressure-driven membrane processes in wastewater treatment and drinking water production. *Environ. Prog.* **2003**, *22*, 46–56. [CrossRef]
6. Introduction. In *The MBR Book*, 2nd ed.; Butterworth-Heinemann: Oxford, UK, 2011; Chapter 1, pp. 1–54. [CrossRef]
7. Abdullah, N.; Rahman, M.A.; Dzarfan Othman, M.H.; Jaafar, J.; Ismail, A.F. Membranes and Membrane Processes: Fundamentals. In *Current Trends and Future Developments on (Bio-) Membranes*; Basile, A., Mozia, S., Molinari, R., Eds.; Elsevier: Amsterdam, The Netherlands, 2018; Chapter 2; pp. 45–70. [CrossRef]
8. Drioli, E.; Ali, A.; Macedonio, F. Membrane distillation: Recent developments and perspectives. *Desalination* **2015**, *356*, 56–84. [CrossRef]
9. Alkhudhiri, A.; Darwish, N.; Hilal, N. Membrane distillation: A comprehensive review. *Desalination* **2012**, *287*, 2–18. [CrossRef]
10. Shirazi, A.; Mahdi, M.; Kargari, A. A review on applications of membrane distillation (MD) process for wastewater treatment. *J. Membr. Sci. Res.* **2015**, *1*, 101–112.
11. Khayet, M.; Wang, R. Mixed Matrix Polytetrafluoroethylene/Polysulfone Electrospun Nanofibrous Membranes for Water Desalination by Membrane Distillation. *ACS Appl. Mater. Interfaces* **2018**, *10*, 24275–24287. [CrossRef]
12. Alkhudhiri, A.; Hilal, N. Air gap membrane distillation: A detailed study of high saline solution. *Desalination* **2017**, *403*, 179–186. [CrossRef]
13. Han, L.; Tan, Y.Z.; Netke, T.; Fane, A.G.; Chew, J.W. Understanding oily wastewater treatment via membrane distillation. *J. Membr. Sci.* **2017**, *539*, 284–294. [CrossRef]
14. Drioli, E.; Ali, A.; Macedonio, F. Membrane operations for process intensification in desalination. *Appl. Sci.* **2017**, *7*, 100. [CrossRef]

15. Bodell, B.R. Distillation of Saline Water Using Silicone Rubber Membrane. U.S. Patents US3361645A, 9 August 1968.

16. Findley, M. Vaporization through porous membranes. *Ind. Eng. Chem. Process Des. Dev.* **1967**, *6*, 226–230. [CrossRef]

17. Gore, D.W. Gore-Tex membrane distillation. In Proceedings of the 10th Annual Conference Water, Honolulu, HI, USA, 25–29 July 1982; pp. 25–29.

18. Khayet, M. Membranes and theoretical modeling of membrane distillation: A review. *Adv. Colloid Interface Sci.* **2011**, *164*, 56–88. [CrossRef] [PubMed]

19. Wang, P.; Chung, T.-S. Recent advances in membrane distillation processes: Membrane development, configuration design and application exploring. *J. Membr. Sci.* **2015**, *474*, 39–56. [CrossRef]

20. Ma, Q.; Ahmadi, A.; Cabassud, C. Direct integration of a vacuum membrane distillation module within a solar collector for small-scale units adapted to seawater desalination in remote places: Design, modeling & evaluation of a flat-plate equipment. *J. Membr. Sci.* **2018**, *564*, 617–633. [CrossRef]

21. Ozbey-Unal, B.; Imer, D.Y.; Keskinler, B.; Koyuncu, I. Boron removal from geothermal water by air gap membrane distillation. *Desalination* **2018**, *433*, 141–150. [CrossRef]

22. Ali, A.; Tufa, R.A.; Macedonio, F.; Curcio, E.; Drioli, E. Membrane technology in renewable-energy-driven desalination. *Renew. Sustain. Energy Rev.* **2018**, *81*, 1–21. [CrossRef]

23. Camacho, L.M.; Dumée, L.; Zhang, J.; Li, J.-D.; Duke, M.; Gomez, J.; Gray, S. Advances in membrane distillation for water desalination and purification applications. *Water* **2013**, *5*, 94–196. [CrossRef]

24. Yang, X.; Wang, R.; Shi, L.; Fane, A.G.; Debowski, M. Performance improvement of PVDF hollow fiber-based membrane distillation process. *J. Membr. Sci.* **2011**, *369*, 437–447. [CrossRef]

25. Wang, P.; Teoh, M.M.; Chung, T.-S. Morphological architecture of dual-layer hollow fiber for membrane distillation with higher desalination performance. *Water Res.* **2011**, *45*, 5489–5500. [CrossRef]

26. Jacob, P.; Laborie, S.; Cabassud, C. Visualizing and evaluating wetting in membrane distillation: New methodology and indicators based on Detection of Dissolved Tracer Intrusion (DDTI). *Desalination* **2018**, *443*, 307–322. [CrossRef]

27. Ali, A.; Tsai, J.-H.; Tung, K.-L.; Drioli, E.; Macedonio, F. Designing and optimization of continuous direct contact membrane distillation process. *Desalination* **2018**, *426*, 97–107. [CrossRef]

28. Sanmartino, J.A.; Khayet, M.; García-Payo, M.C. Desalination by Membrane Distillation. In *Emerging Membrane Technology for Sustainable Water Treatment*; Hankins, N.P., Singh, R., Eds.; Elsevier: Boston, MA, USA, 2016; Chapter 4; pp. 77–109. [CrossRef]

29. Terki, L.; Kujawski, W.; Kujawa, J.; Kurzawa, M.; Filipiak-Szok, A.; Chrzanowska, E.; Khaled, S.; Madani, K. Implementation of osmotic membrane distillation with various hydrophobic porous membranes for concentration of sugars solutions and preservation of the quality of cactus pear juice. *J. Food Eng.* **2018**, *230*, 28–38. [CrossRef]

30. Ding, Z.; Liu, L.; Yu, J.; Ma, R.; Yang, Z. Concentrating the extract of traditional Chinese medicine by direct contact membrane distillation. *J. Membr. Sci.* **2008**, *310*, 539–549. [CrossRef]

31. Khayet, M. Treatment of radioactive wastewater solutions by direct contact membrane distillation using surface modified membranes. *Desalination* **2013**, *321*, 60–66. [CrossRef]

32. Mejia Mendez, D.L.; Castel, C.; Lemaitre, C.; Favre, E. Membrane distillation (MD) processes for water desalination applications. Can dense selfstanding membranes compete with microporous hydrophobic materials? *Chem. Eng. Sci.* **2018**, *188*, 84–96. [CrossRef]

33. Kiss, A.A.; Kattan Readi, O.M. An industrial perspective on membrane distillation processes. *J. Chem. Technol. Biotechnol.* **2018**, *93*, 2047–2055. [CrossRef]

34. Warsinger, D.M.; Swaminathan, J.; Guillen-Burrieza, E.; Arafat, H.A. Scaling and fouling in membrane distillation for desalination applications: A review. *Desalination* **2015**, *356*, 294–313. [CrossRef]

35. Tijing, L.D.; Woo, Y.C.; Choi, J.-S.; Lee, S.; Kim, S.-H.; Shon, H.K. Fouling and its control in membrane distillation—A review. *J. Membr. Sci.* **2015**, *475*, 215–244. [CrossRef]

36. Qtaishat, M.R.; Banat, F. Desalination by solar powered membrane distillation systems. *Desalination* **2013**, *308*, 186–197. [CrossRef]

37. Luo, W.; Phan, H.V.; Li, G.; Hai, F.I.; Price, W.E.; Elimelech, M.; Nghiem, L.D. An osmotic membrane bioreactor–membrane distillation system for simultaneous wastewater reuse and seawater desalination: Performance and implications. *Environ. Sci. Technol.* **2017**, *51*, 14311–14320. [CrossRef]

38. Susanto, H. Towards practical implementations of membrane distillation. *Chem. Eng. Process. Process Intensif.* **2011**, *50*, 139–150. [CrossRef]

39. Curcio, E.; Drioli, E. Membrane distillation and related operations—A review. *Sep. Purif. Rev.* **2005**, *34*, 35–86. [CrossRef]

40. Khayet, M. Membrane distillation. In *Advanced Membrane Technology and Applications*; Elsevier: Amsterdam, The Netherlands, 2008; pp. 297–369.

41. Zhang, J.; Gray, S. Effect of applied pressure on performance of PTFE membrane in DCMD. *J. Membr. Sci.* **2011**, *369*, 514–525. [CrossRef]

42. Liu, F.; Hashim, N.A.; Liu, Y.; Abed, M.M.; Li, K. Progress in the production and modification of PVDF membranes. *J. Membr. Sci.* **2011**, *375*, 1–27. [CrossRef]

43. Martínez, L.; Florido-Díaz, F.; Hernandez, A.; Prádanos, P. Characterisation of three hydrophobic porous membranes used in membrane distillation: Modelling and evaluation of their water vapour permeabilities. *J. Membr. Sci.* **2002**, *203*, 15–27. [CrossRef]

44. Izquierdo-Gil, M.; Garcıa-Payo, M.; Fernández-Pineda, C. Air gap membrane distillation of sucrose aqueous solutions. *J. Membr. Sci.* **1999**, *155*, 291–307. [CrossRef]

45. Alkhudhiri, A.; Hilal, N. 3-Membrane distillation—Principles, applications, configurations, design, and implementation. In *Emerging Technologies for Sustainable Desalination Handbook*; Gude, V.G., Ed.; Butterworth-Heinemann: Oxford, UK, 2018; pp. 55–106. [CrossRef]

46. Zare, S.; Kargari, A. 4–Membrane properties in membrane distillation. In *Emerging Technologies for Sustainable Desalination Handbook*; Gude, V.G., Ed.; Butterworth-Heinemann: Oxford, UK, 2018; pp. 107–456. [CrossRef]

47. Essalhi, M.; Khayet, M. Membrane Distillation (MD). In *Progress in Filtration and Separation*; Elsevier: Amsterdam, The Netherlands, 2015; pp. 61–99.

48. Hassan, A.S.; Fath, H.E. Review and assessment of the newly developed MD for desalination processes. *Desalin. Water Treat.* **2013**, *51*, 574–585. [CrossRef]

49. Li, F.; Huang, J.; Xia, Q.; Lou, M.; Yang, B.; Tian, Q.; Liu, Y. Direct contact membrane distillation for the treatment of industrial dyeing wastewater and characteristic pollutants. *Sep. Purif. Technol.* **2018**, *195*, 83–91. [CrossRef]

50. Martinez, L.; Florido-Diaz, F. Theoretical and experimental studies on desalination using membrane distillation. *Desalination* **2001**, *139*, 373–379. [CrossRef]

51. Wang, K.Y.; Foo, S.W.; Chung, T.-S. Mixed matrix PVDF hollow fiber membranes with nanoscale pores for desalination through direct contact membrane distillation. *Ind. Eng. Chem. Res.* **2009**, *48*, 4474–4483. [CrossRef]

52. Su, M.; Teoh, M.M.; Wang, K.Y.; Su, J.; Chung, T.-S. Effect of inner-layer thermal conductivity on flux enhancement of dual-layer hollow fiber membranes in direct contact membrane distillation. *J. Membr. Sci.* **2010**, *364*, 278–289. [CrossRef]

53. Asadi, R.Z.; Suja, F.; Tarkian, F.; Mashhoon, F.; Rahimi, S.; Jameh, A.A. Solar desalination of Gas Refinery wastewater using membrane distillation process. *Desalination* **2012**, *291*, 56–64. [CrossRef]

54. Alkhudhiri, A.; Darwish, N.; Hilal, N. Produced water treatment: Application of air gap membrane distillation. *Desalination* **2013**, *309*, 46–51. [CrossRef]

55. Summers, E.K.; Arafat, H.A. Energy efficiency comparison of single-stage membrane distillation (MD) desalination cycles in different configurations. *Desalination* **2012**, *290*, 54–66. [CrossRef]

56. Duong, H.C.; Chivas, A.R.; Nelemans, B.; Duke, M.; Gray, S.; Cath, T.Y.; Nghiem, L.D. Treatment of RO brine from CSG produced water by spiral-wound air gap membrane distillation—A pilot study. *Desalination* **2015**, *366*, 121–129. [CrossRef]

57. Khalifa, A.; Lawal, D.; Antar, M.; Khayet, M. Experimental and theoretical investigation on water desalination using air gap membrane distillation. *Desalination* **2015**, *376*, 94–108. [CrossRef]

58. Duong, H.C.; Cooper, P.; Nelemans, B.; Cath, T.Y.; Nghiem, L.D. Evaluating energy consumption of air gap membrane distillation for seawater desalination at pilot scale level. *Sep. Purif. Technol.* **2016**, *166*, 55–62. [CrossRef]

59. Cheng, L.; Zhao, Y.; Li, P.; Li, W.; Wang, F. Comparative study of air gap and permeate gap membrane distillation using internal heat recovery hollow fiber membrane module. *Desalination* **2018**, *426*, 42–49. [CrossRef]

60. Kalla, S.; Upadhyaya, S.; Singh, K. Principles and advancements of air gap membrane distillation. *Rev. Chem. Eng.* **2018**. [CrossRef]
61. Laqbaqbi, M.; Sanmartino, J.A.; Khayet, M.; García-Payo, C.; Chaouch, M. Fouling in membrane distillation, osmotic distillation and osmotic membrane distillation. *Appl. Sci.* **2017**, *7*, 334. [CrossRef]
62. Rivier, C.; García-Payo, M.; Marison, I.; Von Stockar, U. Separation of binary mixtures by thermostatic sweeping gas membrane distillation: I. Theory and simulations. *J. Membr. Sci.* **2002**, *201*, 1–16. [CrossRef]
63. Perfilov, V.; Fila, V.; Sanchez Marcano, J. A general predictive model for sweeping gas membrane distillation. *Desalination* **2018**, *443*, 285–306. [CrossRef]
64. Rahimpour, M.R.; Kazerooni, N.M.; Parhoudeh, M. Water Treatment by Renewable Energy-Driven Membrane Distillation. In *Current Trends and Future Developments on (Bio-) Membranes*; Elsevier: Amsterdam, The Netherlands, 2019; pp. 179–211.
65. Shukla, S.; Méricq, J.; Belleville, M.; Hengl, N.; Benes, N.; Vankelecom, I.; Marcano, J.S. Process intensification by coupling the Joule effect with pervaporation and sweeping gas membrane distillation. *J. Membr. Sci.* **2018**, *545*, 150157. [CrossRef]
66. Moore, S.E.; Mirchandani, S.D.; Karanikola, V.; Nenoff, T.M.; Arnold, R.G.; Sáez, A.E. Process modeling for economic optimization of a solar driven sweeping gas membrane distillation desalination system. *Desalination* **2018**, *437*, 108–120. [CrossRef]
67. Guillén-Burrieza, E.; Blanco, J.; Zaragoza, G.; Alarcón, D.-C.; Palenzuela, P.; Ibarra, M.; Gernjak, W. Experimental analysis of an air gap membrane distillation solar desalination pilot system. *J. Membr. Sci.* **2011**, *379*, 386–396. [CrossRef]
68. Li, L.; Sirkar, K.K. Studies in vacuum membrane distillation with flat membranes. *J. Membr. Sci.* **2017**, *523*, 225–234. [CrossRef]
69. Chen, Q.; Ja, M.K.; Li, Y.; Chua, K. Thermodynamic optimization of a vacuum multi-effect membrane distillation system for liquid desiccant regeneration. *Appl. Energy* **2018**, *230*, 960–973. [CrossRef]
70. Walton, J.; Lu, H.; Turner, C.; Solis, S.; Hein, H. Solar and waste heat desalination by membrane distillation. In *Desalination and Water Purification Research and Development Program Report*; Research and Development Office: El Paso, TX, USA, 2004; p. 20.
71. Xie, Z.; Duong, T.; Hoang, M.; Nguyen, C.; Bolto, B. Ammonia removal by sweep gas membrane distillation. *Water Res.* **2009**, *43*, 1693–1699. [CrossRef]
72. Meindersma, G.W.; Guijt, C.M.; de Haan, A.B. Desalination and water recycling by air gap membrane distillation. *Desalination* **2006**, *187*, 291–301. [CrossRef]
73. Souhaimi, M.K.; Matsuura, T. *Membrane Distillation: Principles and Applications*; Elsevier: Amsterdam, The Netherlands, 2011.
74. Tan, Y.Z.; Han, L.; Chew, N.G.P.; Chow, W.H.; Wang, R.; Chew, J.W. Membrane distillation hybridized with a thermoelectric heat pump for energy-efficient water treatment and space cooling. *Appl. Energy* **2018**, *231*, 1079–1088. [CrossRef]
75. García-Payo, M.; Rivier, C.; Marison, I.; Von Stockar, U. Separation of binary mixtures by thermostatic sweeping gas membrane distillation: II. Experimental results with aqueous formic acid solutions. *J. Membr. Sci.* **2002**, *198*, 197–210. [CrossRef]
76. Bamufleh, H.; Abdelhady, F.; Baaqeel, H.M.; El-Halwagi, M.M. Optimization of multi-effect distillation with brine treatment via membrane distillation and process heat integration. *Desalination* **2017**, *408*, 110–118. [CrossRef]
77. Jansen, A.; Assink, J.; Hanemaaijer, J.; Van Medevoort, J.; Van Sonsbeek, E. Development and pilot testing of full-scale membrane distillation modules for deployment of waste heat. *Desalination* **2013**, *323*, 55–65. [CrossRef]
78. Boutikos, P.; Mohamed, E.S.; Mathioulakis, E.; Belessiotis, V. A theoretical approach of a vacuum multi-effect membrane distillation system. *Desalination* **2017**, *422*, 25–41. [CrossRef]
79. Andrés-Mañas, J.A.; Ruiz-Aguirre, A.; Acién, F.G.; Zaragoza, G. Assessment of a pilot system for seawater desalination based on vacuum multi-effect membrane distillation with enhanced heat recovery. *Desalination* **2018**, *443*, 110–121. [CrossRef]
80. Zhao, K.; Heinzl, W.; Wenzel, M.; Büttner, S.; Bollen, F.; Lange, G.; Heinzl, S.; Sarda, N. Experimental study of the memsys vacuum-multi-effect-membrane-distillation (V-MEMD) module. *Desalination* **2013**, *323*, 150–160. [CrossRef]

81. Heinzl, W.; Büttner, S.; Lange, G. Industrialized modules for MED Desalination with polymer surfaces. *Desalin. Water Treat.* **2012**, *42*, 177–180. [CrossRef]
82. Mohamed, E.S.; Boutikos, P.; Mathioulakis, E.; Belessiotis, V. Experimental evaluation of the performance and energy efficiency of a vacuum multi-effect membrane distillation system. *Desalination* **2017**, *408*, 70–80. [CrossRef]
83. Francis, L.; Ghaffour, N.; Alsaadi, A.A.; Amy, G.L. Material gap membrane distillation: A new design for water vapor flux enhancement. *J. Membr. Sci.* **2013**, *448*, 240–247. [CrossRef]
84. Swaminathan, J.; Chung, H.W.; Warsinger, D.M.; AlMarzooqi, F.A.; Arafat, H.A.; Lienhard V, J.H. Energy efficiency of permeate gap and novel conductive gap membrane distillation. *J. Membr. Sci.* **2016**, *502*, 171–178. [CrossRef]
85. Winter, D.; Koschikowski, J.; Wieghaus, M. Desalination using membrane distillation: Experimental studies on full scale spiral wound modules. *J. Membr. Sci.* **2011**, *375*, 104–112. [CrossRef]
86. Swaminathan, J.; Chung, H.W.; Warsinger, D.M.; Lienhard V, J.H. Membrane distillation model based on heat exchanger theory and configuration comparison. *Appl. Energy* **2016**, *184*, 491–505. [CrossRef]
87. Kujawska, A.; Kujawski, J.K.; Bryjak, M.; Cichosz, M.; Kujawski, W. Removal of volatile organic compounds from aqueous solutions applying thermally driven membrane processes. 2. Air gap membrane distillation. *J. Membr. Sci.* **2016**, *499*, 245–256. [CrossRef]
88. Kalla, S.; Upadhyaya, S.; Singh, K.; Baghel, R. Experimental and mathematical study of air gap membrane distillation for aqueous HCl azeotropic separation. *J. Chem. Technol. Biotechnol.* **2018**. [CrossRef]
89. Quist-Jensen, C.A.; Ali, A.; Mondal, S.; Macedonio, F.; Drioli, E. A study of membrane distillation and crystallization for lithium recovery from high-concentrated aqueous solutions. *J. Membr. Sci.* **2016**, *505*, 167–173. [CrossRef]
90. Al-Obaidani, S.; Curcio, E.; Macedonio, F.; Di Profio, G.; Al-Hinai, H.; Drioli, E. Potential of membrane distillation in seawater desalination: thermal efficiency, sensitivity study and cost estimation. *J. Membr. Sci.* **2008**, *323*, 85–98. [CrossRef]
91. Khalifa, A.; Ahmad, H.; Antar, M.; Laoui, T.; Khayet, M. Experimental and theoretical investigations on water desalination using direct contact membrane distillation. *Desalination* **2017**, *404*, 22–34. [CrossRef]
92. Petrotos, K.B.; Lazarides, H.N. Osmotic concentration of liquid foods. *J. Food Eng.* **2001**, *49*, 201–206. [CrossRef]
93. Alves, V.; Coelhoso, I. Orange juice concentration by osmotic evaporation and membrane distillation: A comparative study. *J. Food Eng.* **2006**, *74*, 125–133. [CrossRef]
94. Kimura, S.; Nakao, S.-I.; Shimatani, S.-I. Transport phenomena in membrane distillation. *J. Membr. Sci.* **1987**, *33*, 285–298. [CrossRef]
95. Hausmann, A.; Sanciolo, P.; Vasiljevic, T.; Kulozik, U.; Duke, M. Performance assessment of membrane distillation for skim milk and whey processing. *J. Dairy Sci.* **2014**, *97*, 56–71. [CrossRef]
96. Criscuoli, A.; Zhong, J.; Figoli, A.; Carnevale, M.; Huang, R.; Drioli, E. Treatment of dye solutions by vacuum membrane distillation. *Water Res.* **2008**, *42*, 5031–5037. [CrossRef] [PubMed]
97. Banat, F.; Al-Asheh, S.; Qtaishat, M. Treatment of waters colored with methylene blue dye by vacuum membrane distillation. *Desalination* **2005**, *174*, 87–96. [CrossRef]
98. Mokhtar, N.M.; Lau, W.J.; Ismail, A.F. Dye wastewater treatment by direct contact membrane distillation using polyvinylidene fluoride hollow fiber membranes. *J. Polym. Eng.* **2015**, *35*, 471–479. [CrossRef]
99. Quist-Jensen, C.A.; Macedonio, F.; Horbez, D.; Drioli, E. Reclamation of sodium sulfate from industrial wastewater by using membrane distillation and membrane crystallization. *Desalination* **2017**, *401*, 112–119. [CrossRef]
100. Kamali, M.; Khodaparast, Z. Review on recent developments on pulp and paper mill wastewater treatment. *Ecotoxicol. Environ. Saf.* **2015**, *114*, 326–342. [CrossRef] [PubMed]
101. Sumathi, S.; Hung, Y.-T. Treatment of pulp and paper mill wastes. In *Waste Treatment in the Process Industries*; Springer: Berlin, Gremnay, 2006; pp. 453–497.
102. Wang, K.Y.; Teoh, M.M.; Nugroho, A.; Chung, T.-S. Integrated forward osmosis–membrane distillation (FO–MD) hybrid system for the concentration of protein solutions. *Chem. Eng. Sci.* **2011**, *66*, 2421–2430. [CrossRef]

103. Gethard, K.; Sae-Khow, O.; Mitra, S. Carbon nanotube enhanced membrane distillation for simultaneous generation of pure water and concentrating pharmaceutical waste. *Sep. Purif. Technol.* **2012**, *90*, 239–245. [CrossRef]

104. Jia, F.; Li, J.; Wang, J.; Sun, Y. Removal of strontium ions from simulated radioactive wastewater by vacuum membrane distillation. *Ann. Nucl. Energy* **2017**, *103*, 363–368. [CrossRef]

105. Liu, H.; Wang, J. Treatment of radioactive wastewater using direct contact membrane distillation. *J. Hazard. Mater.* **2013**, *261*, 307–315. [CrossRef]

106. Reis, B.G.; Araújo, A.L.B.; Amaral, M.C.S.; Ferraz, H.C. Comparison of Nanofiltration and Direct Contact Membrane Distillation as an alternative for gold mining effluent reclamation. *Chem. Eng. Process. Process Intensif.* **2018**, *133*, 24–33. [CrossRef]

107. Woldemariam, D.; Kullab, A.; Khan, E.U.; Martin, A. Recovery of ethanol from scrubber-water by district heat-driven membrane distillation: Industrial-scale technoeconomic study. *Renew. Energy* **2018**, *128*, 484–494. [CrossRef]

108. Onishi, V.C.; Fraga, E.S.; Reyes-Labarta, J.A.; Caballero, J.A. Desalination of shale gas wastewater: Thermal and membrane applications for zero-liquid discharge. In *Emerging Technologies for Sustainable Desalination Handbook*; Elsevier: Amsterdam, The Netherlands, 2018; pp. 399–431.

109. Aminfard, S.; Davidson, F.T.; Webber, M.E. Multi-layered spatial methodology for assessing the technical and economic viability of using renewable energy to power brackish groundwater desalination. *Desalination* **2019**, *450*, 12–20. [CrossRef]

110. Borsani, R.; Rebagliati, S. Fundamentals and costing of MSF desalination plants and comparison with other technologies. *Desalination* **2005**, *182*, 29–37. [CrossRef]

111. Turek, M.; Dydo, P. Hybrid membrane-thermal versus simple membrane systems. *Desalination* **2003**, *157*, 51–56. [CrossRef]

112. Quist-Jensen, C.A.; Macedonio, F.; Drioli, E. Integrated membrane desalination systems with membrane crystallization units for resource recovery: A new approach for mining from the sea. *Crystals* **2016**, *6*, 36. [CrossRef]

113. Dolar, D.; Košutić, K.; Strmecky, T. Hybrid processes for treatment of landfill leachate: Coagulation/UF/NF-RO and adsorption/UF/NF-RO. *Sep. Purif. Technol.* **2016**, *168*, 39–46. [CrossRef]

114. Lokare, O.R.; Tavakkoli, S.; Rodriguez, G.; Khanna, V.; Vidic, R.D. Integrating membrane distillation with waste heat from natural gas compressor stations for produced water treatment in Pennsylvania. *Desalination* **2017**, *413*, 144–153. [CrossRef]

115. Ali, A.; Quist-Jensen, C.A.; Drioli, E.; Macedonio, F. Evaluation of integrated microfiltration and membrane distillation/crystallization processes for produced water treatment. *Desalination* **2018**, *434*, 161–168. [CrossRef]

116. Boukhriss, M.; Khemili, S.; Hamida, M.B.B.; Bacha, H.B. Simulation and experimental study of an AGMD membrane distillation pilot for the desalination of seawater or brackish water with zero liquid discharged. *Heat Mass Transf.* **2018**, 1–11. [CrossRef]

117. Baghbanzadeh, M.; Rana, D.; Lan, C.Q.; Matsuura, T. Zero thermal input membrane distillation, a zero-waste and sustainable solution for freshwater shortage. *Appl. Energy* **2017**, *187*, 910–928. [CrossRef]

118. Huang, X.; Luo, X.; Chen, J.; Yang, Z.; Chen, Y.; Ponce-Ortega, J.M.; El-Halwagi, M.M. Synthesis and dual-objective optimization of industrial combined heat and power plants compromising the water–energy nexus. *Appl. Energy* **2018**, *224*, 448–468. [CrossRef]

119. Moradi, R.; Monfared, S.M.; Amini, Y.; Dastbaz, A. Vacuum enhanced membrane distillation for trace contaminant removal of heavy metals from water by electrospun PVDF/TiO$_2$ hybrid membranes. *Korean J. Chem. Eng.* **2016**, *33*, 2160–2168. [CrossRef]

120. Rácz, G.; Kerker, S.; Schmitz, O.; Schnabel, B.; Kovacs, Z.; Vatai, G.; Ebrahimi, M.; Czermak, P. Experimental determination of liquid entry pressure (LEP) in vacuum membrane distillation for oily wastewaters. *Membr. Water Treat.* **2015**, *6*, 237–249. [CrossRef]

121. Carnevale, M.; Gnisci, E.; Hilal, J.; Criscuoli, A. Direct contact and vacuum membrane distillation application for the olive mill wastewater treatment. *Sep. Purif. Technol.* **2016**, *169*, 121–127. [CrossRef]

122. Mokhtar, N.; Lau, W.; Ismail, A.; Kartohardjono, S.; Lai, S.; Teoh, H. The potential of direct contact membrane distillation for industrial textile wastewater treatment using PVDF-Cloisite 15A nanocomposite membrane. *Chem. Eng. Res. Des.* **2016**, *111*, 284–293. [CrossRef]

123. Wen, X.; Li, F.; Zhao, X. Removal of nuclides and boron from highly saline radioactive wastewater by direct contact membrane distillation. *Desalination* **2016**, *394*, 101–107. [CrossRef]

124. Sardari, K.; Fyfe, P.; Lincicome, D.; Wickramasinghe, S.R. Combined electrocoagulation and membrane distillation for treating high salinity produced waters. *J. Membr. Sci.* **2018**. [CrossRef]

125. Zheng, R.; Chen, Y.; Wang, J.; Song, J.; Li, X.-M.; He, T. Preparation of omniphobic PVDF membrane with hierarchical structure for treating saline oily wastewater using direct contact membrane distillation. *J. Membr. Sci* **2018**, *555*, 197–205. [CrossRef]

126. Cui, Z.; Zhang, Y.; Li, X.; Wang, X.; Drioli, E.; Wang, Z.; Zhao, S. Optimization of novel composite membranes for water and mineral recovery by vacuum membrane distillation. *Desalination* **2018**, *440*, 39–47. [CrossRef]

127. Peydayesh, M.; Kazemi, P.; Bandegi, A.; Mohammadi, T.; Bakhtiari, O. Treatment of bentazon herbicide solutions by vacuum membrane distillation. *J. Water Process Eng.* **2015**, *8*, e17–e22. [CrossRef]

128. Wu, Y.; Kang, Y.; Zhang, L.; Qu, D.; Cheng, X.; Feng, L. Performance and fouling mechanism of direct contact membrane distillation (DCMD) treating fermentation wastewater with high organic concentrations. *J. Environ. Sci.* **2018**, *65*, 253–261. [CrossRef] [PubMed]

129. Naidu, G.; Jeong, S.; Vigneswaran, S.; Hwang, T.-M.; Choi, Y.-J.; Kim, S.-H. A review on fouling of membrane distillation. *Desalin. Water Treat.* **2016**, *57*, 10052–10076. [CrossRef]

130. Bush, J.A.; Vanneste, J.; Cath, T.Y. Comparison of membrane distillation and high-temperature nanofiltration processes for treatment of silica-saturated water. *J. Membr. Sci* **2019**, *570*, 258–269. [CrossRef]

131. Tow, E.W.; Warsinger, D.M.; Trueworthy, A.M.; Swaminathan, J.; Thiel, G.P.; Zubair, S.M.; Myerson, A.S. Comparison of fouling propensity between reverse osmosis, forward osmosis, and membrane distillation. *J. Membr. Sci.* **2018**, *556*, 352–364. [CrossRef]

132. Tang, C.Y.; Chong, T.; Fane, A.G. Colloidal interactions and fouling of NF and RO membranes: A review. *Adv. Colloid Interface Sci.* **2011**, *164*, 126–143. [CrossRef]

133. Shan, H.; Liu, J.; Li, X.; Li, Y.; Tezel, F.H.; Li, B.; Wang, S. Nanocoated amphiphobic membrane for flux enhancement and comprehensive anti-fouling performance in direct contact membrane distillation. *J. Membr. Sci.* **2018**, *567*, 166–180. [CrossRef]

134. Gryta, M. Direct contact membrane distillation with crystallization applied to NaCl solutions. *Chem. Pap. Slovak Acad. Sci.* **2002**, *56*, 14–19.

135. Guillen-Burrieza, E.; Thomas, R.; Mansoor, B.; Johnson, D.; Hilal, N.; Arafat, H. Effect of dry-out on the fouling of PVDF and PTFE membranes under conditions simulating intermittent seawater membrane distillation (SWMD). *J. Membr. Sci.* **2013**, *438*, 126–139. [CrossRef]

136. Gryta, M. Polyphosphates used for membrane scaling inhibition during water desalination by membrane distillation. *Desalination* **2012**, *285*, 170–176. [CrossRef]

137. Nghiem, L.D.; Cath, T. A scaling mitigation approach during direct contact membrane distillation. *Sep. Purif. Technol.* **2011**, *80*, 315–322. [CrossRef]

138. Curcio, E.; Ji, X.; Di Profio, G.; Sulaiman, A.O.; Fontananova, E.; Drioli, E. Membrane distillation operated at high seawater concentration factors: Role of the membrane on $CaCO_3$ scaling in presence of humic acid. *J. Membr. Sci.* **2010**, *346*, 263–269. [CrossRef]

139. Huber, S.A.; Balz, A.; Abert, M.; Pronk, W. Characterisation of aquatic humic and non-humic matter with size-exclusion chromatography – organic carbon detection – organic nitrogen detection (LC-OCD-OND). *Water Res.* **2011**, *45*, 879–885. [CrossRef] [PubMed]

140. Gryta, M.; Grzechulska-Damszel, J.; Markowska, A.; Karakulski, K. The influence of polypropylene degradation on the membrane wettability during membrane distillation. *J. Membr. Sci.* **2009**, *326*, 493–502. [CrossRef]

141. Naidu, G.; Jeong, S.; Vigneswaran, S. Interaction of humic substances on fouling in membrane distillation for seawater desalination. *Chem. Eng. J.* **2015**, *262*, 946–957. [CrossRef]

142. Sutzkover-Gutman, I.; Hasson, D. Feed water pretreatment for desalination plants. *Desalination* **2010**, *264*, 289–296. [CrossRef]

143. Vedavyasan, C.V. Pretreatment trends—An overview. *Desalination* **2007**, *203*, 296–299. [CrossRef]

144. Prihasto, N.; Liu, Q.-F.; Kim, S.-H. Pre-treatment strategies for seawater desalination by reverse osmosis system. *Desalination* **2009**, *249*, 308–316. [CrossRef]

145. Gryta, M. Alkaline scaling in the membrane distillation process. *Desalination* **2008**, *228*, 128–134. [CrossRef]

146. Chen, G.; Yang, X.; Wang, R.; Fane, A.G. Performance enhancement and scaling control with gas bubbling in direct contact membrane distillation. *Desalination* **2013**, *308*, 47–55. [CrossRef]

147. Hickenbottom, K.L.; Cath, T.Y. Sustainable operation of membrane distillation for enhancement of mineral recovery from hypersaline solutions. *J. Membr. Sci.* **2014**, *454*, 426–435. [CrossRef]

148. Xu, J.; Lange, S.; Bartley, J.; Johnson, R. Alginate-coated microporous PTFE membranes for use in the osmotic distillation of oily feeds. *J. Membr. Sci.* **2004**, *240*, 81–89. [CrossRef]

149. Zhang, J.; Song, Z.; Li, B.; Wang, Q.; Wang, S. Fabrication and characterization of superhydrophobic poly (vinylidene fluoride) membrane for direct contact membrane distillation. *Desalination* **2013**, *324*, 1–9. [CrossRef]

150. Lyster, E.; Kim, M.-m.; Au, J.; Cohen, Y. A method for evaluating antiscalant retardation of crystal nucleation and growth on RO membranes. *J. Membr. Sci.* **2010**, *364*, 122–131. [CrossRef]

151. Burgoyne, A.; Vahdati, M.M. Direct Contact Membrane Distillation. *Sep. Sci. Technol.* **2000**, *35*, 1257–1284. [CrossRef]

152. Rezaei, M.; Warsinger, D.M.; Samhaber, W.M. Wetting prevention in membrane distillation through superhydrophobicity and recharging an air layer on the membrane surface. *J. Membr. Sci.* **2017**, *530*, 42–52. [CrossRef]

153. Qtaishat, M.R.; Matsuura, T. 13-Modelling of pore wetting in membrane distillation compared with pervaporation. In *Pervaporation, Vapour Permeation and Membrane Distillation*; Basile, A., Figoli, A., Khayet, M., Eds.; Woodhead Publishing: Oxford, UK, 2015; pp. 385–413. [CrossRef]

154. El-Bourawi, M.S.; Ding, Z.; Ma, R.; Khayet, M. A framework for better understanding membrane distillation separation process. *J. Membr. Sci.* **2006**, *285*, 4–29. [CrossRef]

155. Ge, J.; Peng, Y.; Li, Z.; Chen, P.; Wang, S. Membrane fouling and wetting in a DCMD process for RO brine concentration. *Desalination* **2014**, *344*, 97–107. [CrossRef]

156. Rezaei, M.; Warsinger, D.M.; Duke, M.C.; Matsuura, T.; Samhaber, W.M. Wetting phenomena in membrane distillation: Mechanisms, reversal, and prevention. *Water Res.* **2018**, *139*, 329–352. [CrossRef]

157. Kim, J.; Kwon, H.; Lee, S.; Lee, S.; Hong, S. Membrane distillation (MD) integrated with crystallization (MDC) for shale gas produced water (SGPW) treatment. *Desalination* **2017**, *403*, 172–178. [CrossRef]

158. Guillen-Burrieza, E.; Mavukkandy, M.; Bilad, M.; Arafat, H. Understanding wetting phenomena in membrane distillation and how operational parameters can affect it. *J. Membr. Sci.* **2016**, *515*, 163–174. [CrossRef]

159. Damtie, M.M.; Kim, B.; Woo, Y.C.; Choi, J.-S. Membrane distillation for industrial wastewater treatment: Studying the effects of membrane parameters on the wetting performance. *Chemosphere* **2018**, *206*, 793–801. [CrossRef] [PubMed]

160. Du, X.; Zhang, Z.; Carlson, K.H.; Lee, J.; Tong, T. Membrane fouling and reusability in membrane distillation of shale oil and gas produced water: Effects of membrane surface wettability. *J. Membr. Sci.* **2018**, *567*, 199–208. [CrossRef]

161. Eykens, L.; De Sitter, K.; Dotremont, C.; Pinoy, L.; Van der Bruggen, B. Coating techniques for membrane distillation: An experimental assessment. *Sep. Purif. Technol.* **2018**, *193*, 38–48. [CrossRef]

162. Chen, L.-H.; Chen, Y.-R.; Huang, A.; Chen, C.-H.; Su, D.-Y.; Hsu, C.-C.; Tsai, F.-Y.; Tung, K.-L. Nanostructure depositions on alumina hollow fiber membranes for enhanced wetting resistance during membrane distillation. *J. Membr. Sci.* **2018**, *564*, 227–236. [CrossRef]

163. Eykens, L.; De Sitter, K.; Dotremont, C.; Pinoy, L.; Van der Bruggen, B. Membrane synthesis for membrane distillation: A review. *Sep. Purif. Technol.* **2017**, *182*, 36–51. [CrossRef]

164. Woo, Y.C.; Chen, Y.; Tijing, L.D.; Phuntsho, S.; He, T.; Choi, J.-S.; Kim, S.-H.; Shon, H.K. CF_4 plasma-modified omniphobic electrospun nanofiber membrane for produced water brine treatment by membrane distillation. *J. Membr. Sci.* **2017**, *529*, 234–242.

165. An, X.; Liu, Z.; Hu, Y. Amphiphobic surface modification of electrospun nanofibrous membranes for anti-wetting performance in membrane distillation. *Desalination* **2018**, *432*, 23–31. [CrossRef]

166. Boo, C.; Elimelech, M. Thermal desalination membranes: Carbon nanotubes keep up the heat. *Nat. Nanotechnol.* **2017**, *12*, 501. [CrossRef]

167. Ashoor, B.; Mansour, S.; Giwa, A.; Dufour, V.; Hasan, S. Principles and applications of direct contact membrane distillation (DCMD): A comprehensive review. *Desalination* **2016**, *398*, 222–246. [CrossRef]

168. Lin, S.; Yip, N.Y.; Elimelech, M. Direct contact membrane distillation with heat recovery: Thermodynamic insights from module scale modeling. *J. Membr. Sci.* **2014**, *453*, 498–515. [CrossRef]

169. Long, R.; Lai, X.; Liu, Z.; Liu, W. Direct contact membrane distillation system for waste heat recovery: Modelling and multi-objective optimization. *Energy* **2018**, *148*, 1060–1068. [CrossRef]

170. Lee, J.-G.; Bak, C.-u.; Thu, K.; Ghaffour, N.; Kim, Y.-D. Effect of seawater-coolant feed arrangement in a waste heat driven multi-stage vacuum membrane distillation system. *Sep. Purif. Technol.* **2018**, *9*, 18–20. [CrossRef]

171. Deshmukh, A.; Boo, C.; Karanikola, V.; Lin, S.; Straub, A.P.; Tong, T.; Warsinger, D.M.; Elimelech, M. Membrane distillation at the water-energy nexus: limits, opportunities, and challenges. *Energy Environ. Sci.* **2018**, *11*, 1177–1196. [CrossRef]

172. Swaminathan, J.; Chung, H.W.; Warsinger, D.M. Energy efficiency of membrane distillation up to high salinity: Evaluating critical system size and optimal membrane thickness. *Appl. Energy* **2018**, *211*, 715–734. [CrossRef]

173. Yang, X.; Fane, A.G.; Wang, R. Membrane distillation: Now and future. In *Desalination*; John Wiley & Sons, Inc.: Hoboken, NJ, USA, 2014; pp. 373–424.

174. Kesieme, U.K.; Milne, N.; Aral, H.; Cheng, C.Y.; Duke, M. Economic analysis of desalination technologies in the context of carbon pricing, and opportunities for membrane distillation. *Desalination* **2013**, *323*, 66–74. [CrossRef]

175. Soomro, M.I.; Kim, W.-S. Performance and economic investigations of solar power tower plant integrated with direct contact membrane distillation system. *Energy Convers. Manag.* **2018**, *174*, 626–638. [CrossRef]

176. Tavakkoli, S.; Lokare, O.R.; Vidic, R.D.; Khanna, V. A techno-economic assessment of membrane distillation for treatment of Marcellus shale produced water. *Desalination* **2017**, *416*, 24–34. [CrossRef]

177. Hitsov, I.; Sitter, K.D.; Dotremont, C.; Nopens, I. Economic modelling and model-based process optimization of membrane distillation. *Desalination* **2018**, *436*, 125–143. [CrossRef]

178. Soomro, M.I.; Kim, W.-S. Parabolic-trough plant integrated with direct-contact membrane distillation system: Concept, simulation, performance, and economic evaluation. *Sol. Energy* **2018**, *173*, 348–361. [CrossRef]

chemengineering

MDPI

Article

A Multivariate Statistical Analyses of Membrane Performance in the Clarification of Citrus Press Liquor

René Ruby-Figueroa [1], Monica Nardi [2], Giovanni Sindona [2], Carmela Conidi [3] and Alfredo Cassano [3,*]

[1] Programa Institucional de Fomento a la Investigación, Desarrollo e Innovación, Universidad Tecnológica Metropolitana, Ignacio Valdivieso 2409 P.O. Box 8940577 San Joaquín, Santiago, Chile; rruby@utem.cl
[2] Department of Chemistry, University of Calabria, via P. Bucci, 12/C, I-87030 Rende (CS), Italy; monica.nardi@unical.it (M.N.); giovanni.sindona@unical.it (G.S.)
[3] Institute on Membrane Technology, ITM-CNR, c/o University of Calabria, via P. Bucci, 17/C, I-87030 Rende (CS), Italy; c.conidi@itm.cnr.it
* Correspondence: a.cassano@itm.cnr.it; Tel.: +39-0984-492067; Fax: +39-0984-402103

Received: 19 October 2018; Accepted: 10 January 2019; Published: 17 January 2019

Abstract: The orange press liquor is a by-product of the orange juice production containing bioactive compounds recognized for their beneficial implications in human health. The recovery of these compounds offers new opportunities for the formulation of products of interest in food, pharmaceutical and cosmetic industry. The clarification of orange press liquor by microfiltration (MF) and/or ultrafiltration (UF) processes is a valid approach to remove macromolecules, colloidal particles, and suspended solids from sugars and bioactive compounds. In this work the clarification of orange press liquor was studied by using three flat-sheet polymeric membranes: a MF membrane with a pore size of 0.2 μm and two UF membranes with nominal molecular weight cut-off (MWCO) of 150 and 200 kDa, respectively. The membrane performance, in terms of permeate flux and membrane rejection towards hesperidin and sugars, was studied according to a multivariate analyses approach. In particular, characteristics influencing the performance of the investigated membranes, such as molecular weight cut-off (MWCO), contact angle, membrane thickness, pore size distribution, as well as operating conditions, including temperature, and operating time, were analysed through the partial least square regression (PLSR). The multivariate method revealed crucial information on variables which are relevant to maximize the permeate flux and to minimize the rejection of hesperidin and sugars in the clarification of orange press liquor.

Keywords: microfiltration (MF); ultrafiltration (UF); orange press liquor; clarification; multivariate analysis

1. Introduction

Oranges contribute significantly to the bulk of world's citrus fruit production accounting for more than 50% of the global citrus production. During the marketing year 2015/2016, the global orange production amounted to about 47.06 million metric tons, with Brazil producing 24% of the world total followed by China and India [1].

Although the juice is the main product derived from orange, various by-products are produced during the manufacturing process. The produced wastes consist mainly in wet peels and whole rejected fruits containing 82% of water [2].

Most of the waste residue from commercial juice extractors is shredded, limed, cured, and pressed into press liquors and press cakes which are then processed independently. Press liquors are semisolid wastes containing soluble sugars (sucrose, glucose, and fructose), insoluble carbohydrates, fiber, organic acids, essential oils, flavonoids, and carotenoids [3]. These residues have a considerable amount of organic matter leading to environmental and health problems due to water runoff and uncontrolled fermentation. At the same time, orange peels and pulp contain several bioactive compounds, such as flavonoids and phenolic acids, recognized for their beneficial implications in human health due to their antioxidant activity and free radical scavenging ability [4].

Recent research and development efforts have aimed at converting the potential of wastes into profitable products creating new segments of production and offsetting the disposal costs [5]. Indeed, polyphenolic compounds are used as raw materials in the production of dietary supplements and functional foods, as colouring and flavouring agents in food industries as well as in health and pharmaceutical industries due to their antibacterial, antiviral, anti-inflammatory, antiallergic, and vasodilatory action [6].

Conventional extraction techniques to recover polyphenolic compounds from agro-food waste matrixes usually rely on solid-liquid extraction (SLE) based on the use of volatile organic compounds, such as ethanol, methanol, or acetone solutions as extractants [7]. However, the use of solvents is characterized by serious problems for both consumers and environment due to their toxicity, high volatility, and non-renewable properties.

The growing interest in the biological activity of phenolic compounds has intensified research efforts to develop novel and sustainable procedures for their extraction, separation, and purification in an efficient and environmentally friendly manner without affecting their stability.

Membrane technologies have received great attention in the last years for the recovery of antioxidants from agricultural by-products due to their advantages over conventional methodologies which include mild operating conditions, low energy requirement, no additives, separation efficiency, and easy scale-up [8]. In particular, microfiltration (MF), ultrafiltration (UF), nanofiltration (NF), and reverse osmosis (RO) have been largely investigated, also in sequential design, for the recovery of phenolic compounds from a wide variety of agricultural products and by-products including olive mill wastewaters [9], artichoke wastewaters [10], wine by-products [11], and citrus by-products [12].

An interesting approach to recover and concentrate valuable compounds from orange press liquor is based on the sequential use of membrane operations including ultrafiltration (UF), nanofiltration (NF), and osmotic distillation (OD) [13]. As in the clarification of fruit juices, MF and UF processes allow to remove high molecular weight compounds like cellulose, hemicellulose, cell debris, pectins, and microorganisms from the raw press liquor overcoming typical drawbacks of conventional methods of clarification which include enzymatic treatment (depectinization), cooling, flocculation (gelatin, silica sol, bentonite and diatomaceous earth), decantation, centrifugation, and filtration [14].

These processes separate the flow from the press liquor into a permeate having a total soluble solids content and an acidity level similar to that of the press liquor and a retentate containing suspended solids such as proteins and fibers and high molecular weight carbohydrates, such as cloud pectins.

It is generally recognized that the performance of MF and UF membranes in term of productivity and selectivity is affected by different factors such as membrane characteristics (e.g., pore size, pore size distribution, and contact angle) [15], as well as by operating and fluid-dynamic conditions, including transmembrane pressure, temperature, and feed flowrate [16].

These parameters have to be carefully selected and optimized in order to control concentration polarization and membrane fouling phenomena due to the accumulation of rejected solutes on the membrane surface or within membrane pores.

Generally, the analysis of membrane performance is carried out by using the "one-factor-at-a-time" approach in which each parameter is studied independently of each other. However, it is crucial to take into account the multivariate nature of membrane processes in which the correlation between the variables is usually non-linear, and several factors affect the filtration phenomena simultaneously.

Earlier studies on membrane filtration have shown that the utilization of multivariate analysis extends the information obtained from univariate analysis [17].

In previous studies, the response surface methodology (RSM) approach has been employed to investigate the interaction of different operating conditions, such as transmembrane pressure (TMP), temperature and feed flowrate on permeate flux [18] and the recovery of antioxidant compounds [19] in the clarification of orange press liquor by UF hollow fibre membranes.

Experimental data of permeate flux and fouling index, obtained in optimized operating conditions, resulted in a good agreement with the predicted values of the regression model. The optimized operating conditions to maximize permeate fluxes and the recovery of antioxidant compounds as well as to minimize fouling index were identified.

The present work aimed at investigating the effect of membrane characteristics such as membrane thickness, pore size distribution, contact angle as well as operating conditions, such as temperature and operating time, on the performance of three different flat-sheet MF and UF membranes in terms of permeate flux and rejection of hesperidin and sugars (glucose, fructose, and sucrose) in the treatment of orange press liquor. To accomplish that, the partial least squares regression (PLSR) was used as a multivariate tool, to correlate the membrane characteristics (grouped in an **X** matrix) with membrane performance (grouped in a **Y** matrix).

2. Theory

Partial least squares regression (PLSR), in its simplest form, can be defined as a statistical method for relating two data matrix, **X** and **Y**, to each other by a linear multivariate model [20–23]. The PLSR applications have been reported in three principal areas: quantitative structure-activity relationship (QSAR) modeling, multivariate calibration, and process monitoring and optimization [23].

As a historical note, PLSR or just named PLS approach was originated around 1975 by Herman Wold for the modeling of complex data [23]. PLS can be defined as a multivariate linear regression methodology, based on the decomposition of the data into a set of orthogonal components or latent variables (LVs) [23–26]. It is recognized as a robust method with a robust statistical basis able to analyze data with noisy, collinear, numerous variables and even missing data-points in both the input (**X** matrix) and output (**Y** matrix) data sets. An essential aspect of this technique is that the output data structure guides the decomposition of the input data in a way that the respective orthogonal components explain as much as possible of the covariance between the input and output [27].

As mentioned above, PLS links the input and the output matrices with "new" variables that are estimated as a linear combination of the original variables or their rotation. The following equation gives these new variables called **X**-scores and denoted by t_a ($a = 1, 2, \ldots$ A):

$$t_{ia} = \sum_k W^*_{ka} X_{ik}; \ (\mathbf{T} = \mathbf{XW^*}) \tag{1}$$

where **W** is the weight matrix that relates the **X**-scores with each variable of **X**. On the other hand, the input matrix **X** can be obtained from the linear combination between the **X**-scores **T** and the loading **P** in order to minimize the **X**-residuals **E**:

$$X_{ik} = \sum_a t_{ia} p_{ak} + e_{ik}; \ (\mathbf{X} = \mathbf{TP'} + \mathbf{E}) \tag{2}$$

Then, the output matrix **Y** can be obtained by means of the following equation:

$$Y_{im} = \sum_a c_{ma} t_{ia} + f_{im}; \ (\mathbf{Y} = \mathbf{TC'} + \mathbf{F}) \tag{3}$$

where **C** is the weight matrix that relates the **X**-scores with each variable of **Y**, meanwhile f_{im} represents the deviation between the observed and modeled responses, and comprises the elements of the **Y**-residuals matrix, **F**.

Finally, the multivariate regression model can be obtained combining Equations (1) and (3):

$$Y_{im}\sum_a c_{ma} \sum_k w^*_{ka} x_{ik} + f_{im} = \sum_k b_{mk} x_{ik} + f_{im}; \ (\mathbf{Y} = \mathbf{XW^*C'} + \mathbf{F} = \mathbf{XB} + \mathbf{F}) \tag{4}$$

The PLS regression coefficients, b_{mk} (**B**) can be written as:

$$b_{mk} = \sum_a c_{ma} w^*_{ka}; \ (\mathbf{B} = \mathbf{W^*C'}) \tag{5}$$

The line obtained by linear regression of that swarm of data points, in the direction of maximum variance, is the first latent variable or just factor. In other words, it captures the main trend in the data set. Then, another linear regression is performed in the second direction of maximum variance, but keeping in mind that this direction should be orthogonal to the first. This corresponds to the second factor. The remaining factors are obtained accordingly [27].

3. Materials and Methods

3.1. Feed Solution

Citrus press liquor was supplied by Gioia Succhi Srl (Rosarno, Reggio Calabria, Italy). Liquors were left overnight at room temperature to let the majority of the cloud particles settle out. Partially clear liquor was recovered by filtration with a nylon cloth. The physico-chemical characteristics of the resulting liquor are reported in Table 1.

Table 1. Physico-chemical characteristics of orange press liquor.

Hesperidin (mg/L)		159.60 ± 14.42
Glucose (mg/mL)		14.69 ± 0.19
Fructose (mg/mL)		20.48 ± 0.11
Sucrose (mg/mL)		2.11 ± 0.13
Total soluble solids (g/100 g)		8.6 ± 0.1
Solid content, after lyophilisation (g/100 mL)		4.94 ± 0.04
Density (kg/L)		1.02835 ± 0.0005
pH		3.58 ± 0.03
Viscosity (cp)	15 °C	1.60 ± 0.02
	25 °C	1.45 ± 0.01
	35 °C	1.31 ± 0.03

3.2. MF-UF Equipment and Procedures

MF and UF experiments were performed by using a laboratory bench plant (Figure 1) equipped with a stainless steel cell suitable to contain a flat-sheet membrane with a diameter of 47.2 mm. Experimental runs were performed by using three polyvinylidenfluoride (PVDF) flat-sheet membranes supplied by Microdyn-Nadir GmbH (Wiesbaden, Germany). Properties of selected membranes are reported in Table 2. Experimental runs were performed according to the total recycle (TR) configuration in which both permeate and retentate streams were continuously recycled back to the feed tank. This configuration ensured a steady-state in the volume and composition of the feed. In order to evaluate the effect of feed concentration on the membrane performance, experiments were also performed according to the batch concentration (BC) configuration in which the permeate stream was continuously removed. In both configurations operating conditions such as transmembrane pressure (TMP), temperature and feed flowrate were fixed at 1 bar, 26.0 ± 1.0 °C and 185 L/h, respectively. Each run was stopped after 180 min of operation. Experimental runs were performed in triplicate. Permeate flux data were expressed as average value $\pm$ SD.

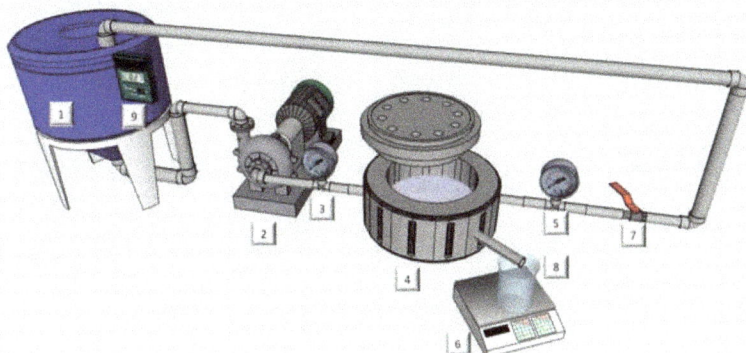

Figure 1. Schematic diagram of the experimental set-up: (1) feed tank; (2) feed pump; (3,5) pressure gauges; (4) flat-sheet cell; (6) digital balance; (7) retentate valve; (8) permeate tank; and (9) thermometer.

Table 2. Characteristics of selected membranes.

Membrane Type	MV020T	UV150T	FMU6R2
Membrane process	MF	UF	UF
Membrane configuration	flat-sheet	flat-sheet	flat-sheet
Membrane material	PVDF	PVDF	PVDF
pH range	2–11	2–11	2–11
Processing temperature (°C)	5–95	5–95	5–95
Thickness (mm)	0.188 ± 0.005 [b]	0.212 ± 0.004 [b]	0.190 ± 0.003 [b]
MWCO (kDa)	-	150 [a]	200 [a]
Pore size (µm)	0.2 [a]	-	-
Maximum pore size distribution (frequency, %)	93.21 [b]	79.77 [b]	40.47 [b]
Diameter at maximum pore size distribution (µm)	0.488 ± 0.048 [b]	0.195 ± 0.019 [b]	0.212 ± 0.021 [b]

MWCO, molecular weight cut-off; PVDF, polyvinylidenfluoride (PVDF). [a] data from the manufacturer; [b] experimental data.

The permeate flux (J) was determined by weighing the amount of permeate with a digital balance and calculated according to the following equation:

$$J = \frac{W_p}{t \cdot A_p} \tag{6}$$

where W_p is the permeate weight collected during the time interval t and A_p is the membrane surface area of permeation. The mass of permeate collected was measured with an accuracy of ± 0.1 g every 5 min.

3.3. Determination of Sugars

The quantitative determination of glucose, fructose and sucrose was carried out by an HPLC system (Agilent Technologies, Palo Alto, CA, USA) equipped with a Luna reverse phase C18 column (5 µ, 100 Å, 250 × 4.6 mm i.d. from Phenomenex (Torrance, CA, USA), an isocratic pump (model series 1100) and a refractive index detector (Series 200a).

Isocratic elution was used at a flow rate of 1 mL/min with two solvents: Solvent A, water/acetic acid (0.1% *v/v*), 80%, and Solvent B, methanol, 20%.

For each reference sugar, a set of calibration standards using stock and working reference standard solutions were prepared. Glucose, fructose, and sucrose were purchased from Sigma-Aldrich (Milan, Italy). Sugar standards were dried at 60 °C in a vacuum oven overnight and dissolved in 50% methanol (injection solvent). The resultant solutions were filtered using a syringe filter and injected into HPLC.

The injection volume was 20 µL. The peak areas in the chromatograms were plotted against calibration curves obtained from standard solutions (external standard method), in a concentration range of 0.5–2 mg/mL for each compound. Results were expressed as mean ± SD of three independent determinations.

3.4. Determination of Hesperidin

The quantitative determination of hesperidin was carried out by an HPLC system (Shimadzu LC-20AB, Kyoto, Japan) equipped with a binary pump, autosampler and a UV/vis detector (SPD-20A), monitored at 284 nm and 360 nm. Samples were centrifuged before injection. The column used was a Discovery C18 (25 cm × 4.6 mm, 5 µm from Supelco, Bellefonte, PA, USA). The mobile phase consisted of two solvents: Solvent A, water/phosphoric acid (0.1% v/v) and Solvent B, acetonitrile. Phenolic compounds were eluted under the following conditions: 1 mL/min flow rate and ambient temperature; gradient conditions from 0% to 5% B in 0.01 min, from 5% to 10% B in 19.9 min, from 10% to 20% B in 20 min, from 20% to 25% B in 20 min, from 25% to 35% B in 20 min, from 35% to 60% B in 15 min, from 60% to 5% B in 3 min, followed by washing and reconditioning of the column. The identification of hesperidin was obtained comparing the retention time by using authentic standard. Hesperidin was from Sigma-Aldrich (Milan, Italy). Results were expressed as mean ± SD of three independent determinations.

3.5. Pore size and Pore Size Distribution Measurement

Membranes pore size and pore size distribution were determined by using a PMI Capillary Flow porometer (Porous Materials Inc., Ithaca, NY, USA) according to the bubble point method [28,29]. A porewick solution (surface tension 16 dynes/cm) was used as a wetting liquid. Fully wetted samples were sealed in the cell and measurements were carried out by the wet up/dry down method using the software Capwin (Porous Materials Inc., USA). Data were processed and exported as an Excel file by the software Caprep (Porous Materials Inc., USA).

3.6. Thickness and Contact Angle Measurement

The thickness of each membrane was determined by a multiple-point measurement, using a digital micrometre Mahr 40E (Mahr GmbH, Esslingen, Germany). Contact angle measurement were carried out by using the sessile drop method with a CAM 200 contact angle meter (KSV instrument LTD, Helsinki, Finland). The droplets were deposited on the membrane surface by using a micro-syringe with automatic dispenser, while the images were captured by a digital camera allowing apparent static contact angles to be measured at different time. An average of 20 readings was obtained for each specimen and the respective mean value was calculated.

3.7. Data Analysis

3.7.1. Pre-Processing

Data were initially organized into dataset X-matrix (n × k) which is composed of 114 observations and five factors or predictors such as membrane thickness, diameter at maximum pore size distribution, contact angle, operating time, and temperature. On the other hand, Y-matrix (n × m), also called response, was composed of 114 observations and five responses: permeate flux and rejection towards hesperidin, glucose, fructose, and sucrose.

In any analytical application, data are usually processed before using PLSR. In this work, in which factors and responses are discrete variables, data were pre-processed in order to obtain the maximum information from the dataset. In general, pre-processing is but a minor modification of the dataset, with the aim of minimizing the impact from extraneous noise and also putting each variable both on an equal level with an equal scaling allowing to participate equally in the data modeling process [30]. Results of projection methods, such as PLSR, depend on the scaling of the data. With an appropriate scaling, one can focus the model on more important Y-variables, and use the experience to increase the weights of more informative X-variables [23,31]. In our case, the absence of knowledge about the relative importance of the variables and the fact that the factors and responses are in different units have forced to probe different techniques such as normalize and moving average.

3.7.2. Number PLSR Components or Factors and Model Validation

In any empirical modeling, it is essential to determine the real complexity of the model. Considering numerous and correlated X-variables, a substantial risk exists for "over-fitting" that means a well-fitting model with little predictive power [23]. Cross-validation (CV) is a practical and reliable way to test the predictive significance [20,32–34]. Basically, in CV the data are divided into groups followed by the development of parallel models that are evaluated with the differences between observed and predicted Y-values. In the evaluation, the observations are kept out of the developed model while the response values (Y) are predicted and compared with the observed values. The procedure is repeated several times until every observation has been kept out. The sum of squares of these differences is computed and collected from all the parallel models to form the predictive residual sum of squares (PRESS), which estimates the predictive ability of the model [23]. The ability of the model can be summarized using the R^2 of the calibration and validation set, the root mean square error of calibration and validation (RMSE), the standard error of calibration and validation (SE is similar to RMSE except it is corrected for the bias) and the bias which is the mean value over all points that either lie systematically above (or below) the regression line (a value close to zero indicate a random distribution of point about the regression line).

All the statistical computations were performed using Unscrambler 10.4.1 software (CAMO AS, Oslo, Norway).

4. Results and Discussion

4.1. Membrane Characteristics

Measurements performed to characterize the selected membranes are shown in Table 2. The membranes used in this study, made of PVDF, are basically hydrophilic membranes with contact angle values lower than 90°. The FMU6R2 membrane showed the most hydrophilic surface, followed by UV150T and MV020T membranes. Thickness measurements showed no significant differences between the membranes studied. The main differences between the selected membranes are related with the pore size. The measurements carried out to evaluate the maximum pore size for each membrane showed differences with the data reported by the manufactures. In addition, significant differences related to the maximum pore size distribution were found. In particular, the FMU6R2 membrane showed the minimum frequency (40%); this means that only 40% of the membrane surface has a pore size of 0.2 μm, and the rest of surface could present higher or lower values. On the other hand, MV020T and UV150T membranes showed higher frequency of distribution: therefore, these membranes are more homogeneous in their pore size distributions. These factors will be strongly related with the type of fouling produced and, therefore, with the membrane performance in terms of permeate flux and rejection towards hesperidin and sugars.

4.2. Time Evolution of Permeate Flux

Figure 2 shows the time evolution of permeate flux under the TR configuration. For all selected membranes the permeate flux reduces constantly due to concentration polarization and fouling phenomena until to reach a uniform rate known as steady-state. In particular, the MV020T membrane and the FMU6R2 membrane showed a quite similar flux decay (35.6 and 31.6%, respectively); for the UV150T membrane the flux decay was of about 41.4%. These effects could be attributed to the type of fouling produced during the treatment of the orange press liquor. As expected, the MF membrane, with larger pore size, exhibited highest permeate flux values in comparison with UF membranes.

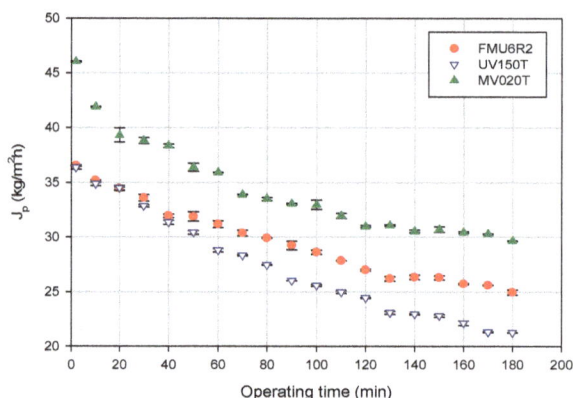

Figure 2. Time course of permeate flux for selected membranes under total recycle configuration. Operating conditions: TMP, 1 bar; temperature, 26.0 ± 1.0 °C; feed flowrate, 185 L/h.

In the BC configuration the permeate stream is continuously removed from the system, while the retentate stream is recycled back to the feed reservoir leading to an increase of the feed concentration during the filtration process. The increased feed concentration results in a more severe concentration polarization and, consequently, in a more pronounced flux decline in comparison with the TC configuration (Figure 3). In these conditions the MF membrane with larger pores showed the maximum flux decay (51.4%), followed by FMU6R2 and UV150T membranes with flux decay values of 38.4% and 36.1%, respectively.

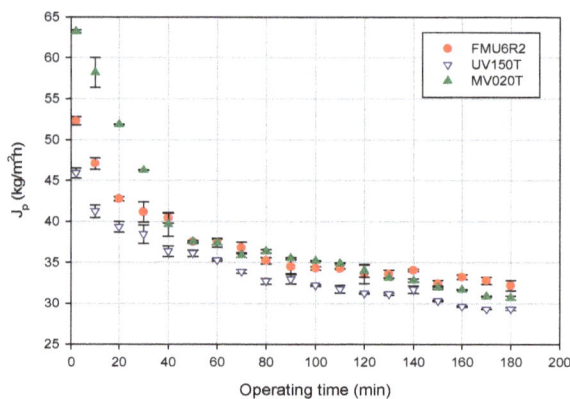

Figure 3. Time course of permeate flux for selected membranes under batch concentration configuration. Operating conditions: TMP, 1 bar; temperature, 26.0 ± 2.0 °C; and feed flowrate, 185 L/h.

4.3. Data Analyses

The data were pre-processed by using several normalization tools and moving average. Table 3 shows the results obtained for the PLSR models, all of them with four components of factors. The pre-processing of area normalization, followed by moving average, was the once which obtained the highest values of R^2 for the calibration and validation, as well as the minimum values of RSME, SE, and bias for all the responses studied. This PLSR model with four factors can explain the 95.64% of the total variance. Even though, there are differences in the capacity of prediction of the PLSR model for each response variable in all the cases the prediction was higher than 91% (R^2), as shown in Table 3.

Table 3. Comparison of various pre-processing methods for the PLSR modeling. Pre-processing: A: Normalize (area normalization) and moving average; B: Normalize (unit vector normalization) and moving average; C: Normalize (mean normalization) and moving average (Cal: Calibration data set; Val: Validation data set).

Response	Parameters	Pre-Processing							
		None		A		B		C	
		Cal	Val	Cal	Val	Cal	Val	Cal	Val
Permeate flux	Slope	0.596	0.576	0.966	0.959	0.978	0.973	0.966	0.959
	R^2	0.596	0.564	0.966	0.962	0.978	0.975	0.966	0.962
	RMSE (C,P)	4.339	4.521	0.003	0.003	0.007	0.008	0.031	0.033
	SE (C,P)	4.358	4.541	0.003	0.003	0.007	0.008	0.031	0.033
	Bias	0	−0.001	0	-7.4×10^{-5}	0	-7.9×10^{-5}	0	-3.4×10^{-4}
Hesperidin	Slope	0.864	0.859	0.963	0.957	0.962	0.961	0.963	0.956
	R^2	0.864	0.852	0.963	0.958	0.962	0.959	0.963	0.958
	RMSE (C,P)	3.906	4.076	0.003	0.004	0.011	0.011	0.035	0.038
	SE (C,P)	3.923	4.094	0.003	0.004	0.011	0.011	0.036	0.038
	Bias	0	−0.032	0	-7.1×10^{-5}	0	4.9×10^{-5}	0	-4.4×10^{-4}
Glucose	Slope	0.280	0.257	0.925	0.921	0.899	0.899	0.925	0.921
	R^2	0.280	0.245	0.925	0.917	0.899	0.894	0.925	0.919
	RMSE (C,P)	7.554	7.853	0.005	0.005	0.018	0.019	0.051	0.054
	SE (C,P)	7.587	7.887	0.005	0.005	0.018	0.019	0.051	0.054
	Bias	0	−0.057	0	-8.9×10^{-5}	0	1.3×10^{-4}	0	−0.0006
Fructose	Slope	0.309	0.286	0.978	0.981	0.675	0.676	0.978	0.981
	R^2	0.309	0.274	0.978	0.975	0.675	0.641	0.978	0.975
	RMSE (C,P)	7.133	7.413	0.003	0.003	0.037	0.039	0.031	0.033
	SE (C,P)	7.164	7.445	0.003	0.003	0.038	0.039	0.031	0.033
	Bias	0	−0.060	0	7.4×10^{-5}	0	0.0006	0	3.4×10^{-4}
Sucrose	Slope	0.052	0.022	0.951	0.9546	0.636	0.637	0.951	0.954
	R^2	0.052	0.001	0.951	0.943	0.636	0.595	0.951	0.946
	RMSE (C,P)	18.465	19.074	0.005	0.006	0.044	0.047	0.053	0.057
	SE (C,P)	18.547	19.159	0.005	0.006	0.044	0.047	0.053	0.057
	Bias	0	−0.051	0	1.1×10^{-4}	0	0.0008	0	0.0006

Figure 4 shows the analysis of the presence of outliers which were carried out by the use of Hotelling T^2 statistic, a multivariate generalization of the student *t*-test [35]. In this figure, several points can be appreciated in the regions 1, 2, and 3. They represent samples similar to the majority of the calibration population, samples which fit the model but are extreme in properties and samples which differ from the average model population, respectively. On the other hand, samples which are different and extreme, those considered as outliers are placed in the region 4. Thus, none of the data was removed for the PLSR modelling.

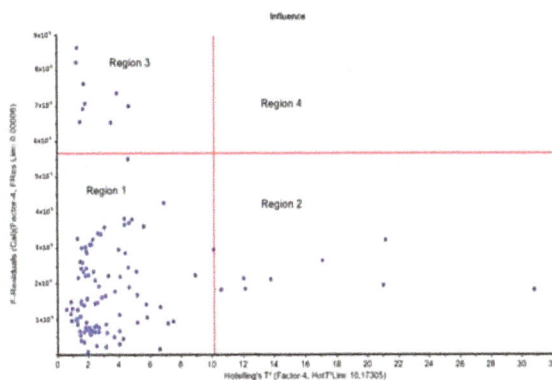

Figure 4. Influence plot with Hotelling's T^2 statistic.

The PLSR scores-plot shown in Figure 4 was used to evaluate the relationship between the samples. Factors 1 and 2, including 100% of the X-matrix data and explaining the 83% of the variability in the Y-matrix, demonstrate that there are differences in the membranes studied and can be grouped according to the tested membrane; this means that each membrane is characterized by specific parameters which discriminate it from each other leading to a specific performance. In particular, the FMU6R2 membrane showed similarities, and it is grouped clearly as a cluster, as well as UV150T membrane placed in the negative sector of factor 2 (Figure 5a). The MV020T membrane has not grouped, and it is placed in the positive and negative part of factor 2. On the other hand, regarding the type of process, the score-plot (Figure 5b) showed a grouping between the TR and BC configuration.

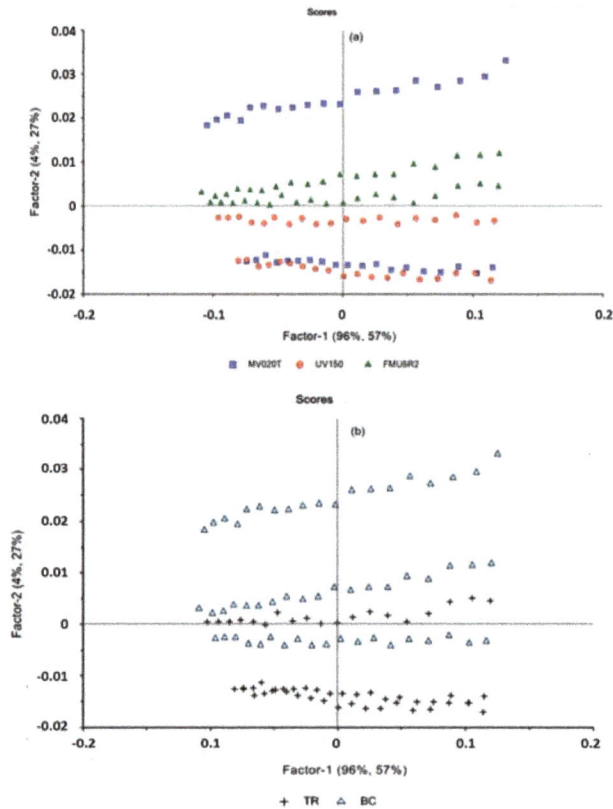

Figure 5. PLS Score plot for the two principal factors. (**a**) Measurements griped by membrane studied, and (**b**) measurements by type of processing.

This group is not clearly observed for the FMU6R2 membrane placed in the origin of factor 2: for this membrane TRC and BC are not grouped. These results highlight the differences not only between the membrane characteristics but also between the type of configuration in which the principal difference is related to the increase of feed concentration which produces a variance in the type of fouling and, consequently, in the membrane performance.

The correlation among all membrane characteristics and operating conditions with the responses variables used to evaluate the membrane performance is illustrated in Figure 6. In this figure differences in the influence of membrane characteristics and operating conditions on the permeate flux and rejection of hesperidin and sugars can be appreciated. In particular, operating time and thickness play a significant role (they are far away from the responses) on the permeate flux and rejections: this means that higher values of thickness and operating time produce a lower value of permeate flux and rejection towards hesperidin and sugars. In this regard, it is well known that an increase in membrane thickness produces an additional resistance to the mass transfer across the membrane. Thus membranes with lower values of membrane thickness are preferred. On the other hand, higher operating times are related to a progressive membrane fouling leading to an increasing of membrane resistance. The decrease in the membrane rejection is related to the concentration polarization phenomena which produces an increase in the particle concentration at the membrane surface where the difference in the chemical potential produces a diffusion of hesperidin and sugars with a decreasing of membrane rejection.

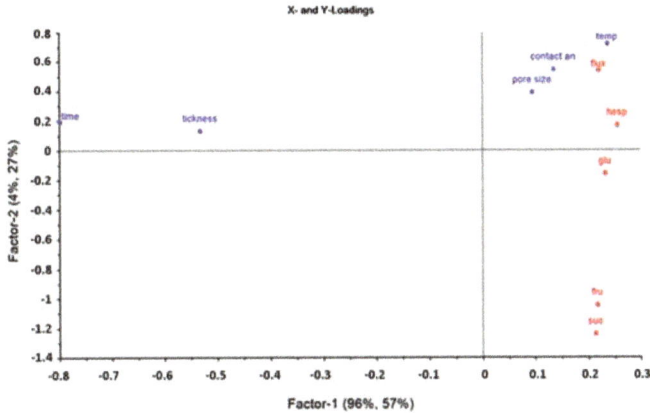

Figure 6. PLS loading plot for the two principal factors.

The loading-plot also shows the positive correlation between temperature, contact angle, and pore size distribution with permeate flux and rejection of hesperidin since they are located in the positive quadrant of factors 1 and 2. Even though these variables have presented less importance in the model, their influences in the responses should not be neglected. According to the film model [36], an increase in temperature enhances permeate flux due to an increase of the mass-transfer coefficient. An increasing in MWCO produces an increase in the rejection towards hesperidin due to the type of fouling produced. In particular, in membranes with larger pores, such as MF membranes, a complete pore blocking or a partial pore blocking is the dominant fouling mechanism which produces a decrease in the pore size and a consequent increase in the rejection as is shown in Figure 7. The physical blockage of the pores also produces a more significant flux decline in comparison with membranes having tight pores. Similar results were obtained by Lin et al. [37] which evaluated the effects of dissolved organic matter retention and membrane pore size on membrane fouling and flux decline.

By referring to the sugars rejection, it is appreciated in factor 2 of the loading-plot that glucose, fructose, and sucrose are negatively related to pore size, contact angle, and temperature. According to results obtained by Jiraratananon and Chanachai [38] in the clarification of passion fruit juice by UF membranes, the operating temperature enhances the back diffusion of solutes into the bulk solution reducing the thickness of the concentration polarization layer. Fructose, glucose, and sucrose rejections showed a similar behavior because are closer in the negative quadrant of factors 1 and 2 in the loading plot.

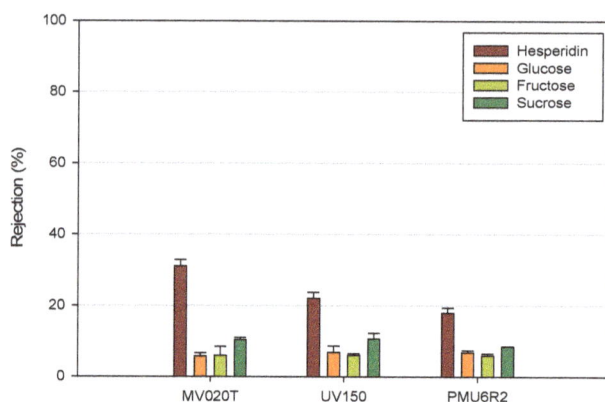

Figure 7. Rejections of sugars and hesperidin for selected membranes under batch concentration configuration. Operating conditions: TMP, 1 bar; temperature, 26.0 ± 1.0 °C; and feed flowrate, 185 L/h.

The PLSR model built after data pre-treatment including four factors is suitable to predict the response variables by correlation of membrane characteristics and operating conditions. Table 3 shows that the model fits well the experimental data with R^2 values of 96.2, 95.8, 91.7, 97.5, and 94.3 for permeate flux, hesperidin, glucose, fructose, and sucrose rejection, respectively. The obtained model can be used to predict the permeate flux, as well as hesperidin and sugars rejections, by using input data such as contact angle, membrane thickness, pore size distribution, as well as operating conditions, such as temperature and process time. The model is consistent with the knowledge obtained in early studies and supplies new information concerning membrane filtration in citrus juice processing.

5. Conclusions

Orange press liquor was clarified by using three flat-sheet MF and UF polymeric membranes in both total recycle and batch concentration configuration. A multivariate analyses approach was used to study the relationship between membrane characteristics and operating conditions and membrane performance in terms of permeate flux and membrane rejection towards hesperidin and sugars (glucose, fructose, and sucrose). In particular, the partial least squares regression (PLSR) model was used in order to predict the response variables by using input data such as contact angle, membrane thickness, pore size distribution as well as operating conditions, such as temperature and process time.

The model well fitted the experimental data with R^2 values of 96.2, 95.8, 91.7, 97.5, and 94.3 for permeate flux, hesperidin, glucose, fructose, and sucrose rejection, respectively. Therefore, the capacity of prediction of response variables resulted higher than 91.7%.

The obtained results indicated that the multivariate method appears as an efficient tool in the examination of experimental results and reveals crucial information on which variables are relevant to maximize the permeate flux and to minimize the rejection of hesperidin and sugars in the clarification of orange press liquor, so maximizing the productivity of the process and the recovery of target compounds in the permeate stream.

Author Contributions: Conceptualization: A.C. and R.R.-F.; methodology: A.C., C.C., R.R.-F. and G.S.; software: R.R.-F.; analytical measurements: M.N.; data elaboration: R.R.-F.; writing—original draft preparation: A.C. and R.R.-F.; supervision: A.C.

Funding: This research received no external funding.

Acknowledgments: The authors wish to thank Silvia Simone from ITM-CNR for her valuable contribution in the characterization of flat-sheet membranes used in the present work.

Conflicts of Interest: The authors declare no conflict of interest.

References

1. Statista. The Statistics Portal. Available online: https://www.statista.com/statistics/577398/world-orange-production (accessed on 12 October 2018).
2. Goodrich, R.M.; Braddock, R.J. Major By-Products of the Florida Citrus Processing Industry, Document FSHN05–22, Series of the Food Science and Human Nutrition Department, Florida Cooperative Extension Service, Institute of Food and Agricultural Sciences, University of Florida, Original Publication Date October 2004. Available online: Ufdcimages.uflib.ufl.edu/IR/00/00/20/62/00001/FS10700.pdf (accessed on 12 October 2018).
3. Garcia-Castello, E.M.; McCutcheon, J.R. Dewatering press liquor derived from orange production by forward osmosis. *J. Membr. Sci.* **2011**, *372*, 97–101. [CrossRef]
4. Imeh, U.; Khokhar, S. Distribution of conjugated and free phenols in fruits: Antioxidant activity and cultivar variations. *J. Agric. Food Chem.* **2002**, *50*, 6301–6306. [CrossRef] [PubMed]
5. Laufenberg, G.; Kunz, B.; Nystroem, M. Transformation of vegetable waste into value added products. *Bioresour. Technol.* **2003**, *87*, 167–198. [CrossRef]
6. Marín, F.R.; Martínez, M.; Uribesalgo, T.; Castillo, S.; Frutos, M.J. Changes in nutraceutical composition of lemon juices according to different industrial extraction systems. *Food. Chem.* **2002**, *78*, 319–324. [CrossRef]
7. Librán, C.; Mayor, L.; Garcia-Castello, E.; Vidal-Brotons, D. Polyphenol extraction from grape wastes: Solvent and pH effect. *Agric. Sci.* **2013**, *4*, 56–62. [CrossRef]
8. Galanakis, C.M. Recovery of high added-value components from food wastes: Conventional, emerging technologies and commercialized applications. *Trends Food Sci. Technol.* **2012**, *26*, 68–87. [CrossRef]
9. Cassano, A.; Conidi, C.; Giorno, L.; Drioli, E. Fractionation of olive mill wastewaters by membrane separation techniques. *J. Hazard. Mater.* **2013**, *248–249*, 185–193. [CrossRef] [PubMed]
10. Conidi, C.; Cassano, A.; Garcia-Castello, E. Valorization of artichoke wastewaters by integrated membrane process. *Water Res.* **2014**, *48*, 363–374. [CrossRef] [PubMed]
11. Giacobbo, A.; Bernardes, A.M.; de Pinho, M.N. Sequential pressure driven membrane operations to recover and fractionate polyphenols and polysaccharides from second racking wine lees. *Sep. Purif. Technol.* **2017**, *173*, 49–54. [CrossRef]
12. Conidi, C.; Cassano, A.; Drioli, E. Recovery of phenolic compounds from orange press liquor by nanofiltration. *Food Bioprod. Process.* **2012**, *90*, 867–874. [CrossRef]
13. Cassano, A.; Conidi, C.; Ruby-Figueroa, R. Recovery of flavonoids from orange press liquor by an integrated membrane process. *Membranes* **2014**, *4*, 509–524. [CrossRef]
14. Rai, P.; Rai, C.; Majumdar, G.C.; DasGupta, S.; De, S. Storage study of ultrafiltered mosambi (*Citrus sinensis* L. Osbeck) juice. *J. Food Process Preserv.* **2008**, *32*, 923–934. [CrossRef]
15. Alsalhy, Q.; Algebory, S.; Alwan, G.M.; Simone, S.; Figoli, A.; Drioli, E. Hollow fiber ultrafiltration membranes from poly(vinylchloride): Preparation, morphologies, and properties. *Sep. Sci. Technol.* **2011**, *46*, 2199–2210. [CrossRef]
16. Girard, B.; Fukumoto, L.R.; Koseoglu, S.S. Membrane processing of fruit juices and beverages: A review. *Crit. Rev. Biotechnol.* **2000**, *20*, 109–175. [CrossRef] [PubMed]
17. Kallioinen, M.; Reinikainen, S.P.; Nuortila-Jokinen, J.; Mänttäri, M.; Sutela, T.; Nurminen, P. Chemometrical approach in studies of membrane capacity in pulp and paper mill application. *Desalination* **2005**, *175*, 87–95. [CrossRef]
18. Ruby-Figueroa, R.; Cassano, A.; Drioli, E. Ultrafiltration of orange press liquor: Optimization for permeate flux and fouling index by response surface methodology. *Sep. Purif. Technol.* **2011**, *80*, 1–10. [CrossRef]
19. Ruby-Figueroa, R.; Cassano, A.; Drioli, E. Ultrafiltration of orange press liquor: Optimization of operating conditions for the recovery of antioxidant compounds by response surface methodology. *Sep. Purif. Technol.* **2012**, *98*, 255–261. [CrossRef]
20. Höskuldsson, A. *Prediction Methods in Science and Technology*; Thor Publishing: Copenhagen, Denmark, 1996.
21. Wold, S.; Ruhe, A.; Wold, H.; Dunn, W.J. The collinearity problem in linear regression, The partial least squares approach to generalized inverses. *SIAM J. Sci. Stat. Comput.* **1984**, *5*, 735–743. [CrossRef]
22. Wold, S.; Sjöström, M.; Eriksson, L. PLS in chemistry. In *The Encyclopedia of Computational Chemistry*; Schleyer, P.V.R., Allinger, N.L., Clerk, T., Gasteiger, J., Kollman, P.A., Schaefer, H.F., III, Schreiner, P.R., Eds.; John Wiley & Sons: Chichester, UK, 1999; pp. 2006–2020.

23. Wold, S.; Sjöström, M.; Eriksson, L. PLS-Regression: A basic tool of chemometrics. *J. Chemom.* **2001**, *58*, 109–130. [CrossRef]
24. Kourti, T.; MacGregor, J.F. Process analysis, monitoring and diagnosis, using multivariate projection methods. *Chemom. Intell. Lab. Syst.* **1995**, *28*, 3–21. [CrossRef]
25. MacGregor, J.F.; Yu, H.; Muñoz, S.G.; Flores-Cerrillo, J. Data-based latent variable methods for process analysis, monitoring and control. *Comput. Chem. Eng.* **2005**, *29*, 1217–1223. [CrossRef]
26. Metsämuuronena, S.; Reinikainenb, S.; Nyström, M. Analysis of protein filtration data by PLS regression. *Desalination* **2002**, *149*, 453–458. [CrossRef]
27. Santos, J.L.C.; Hidalgo, A.M.; Oliveira, R.; Velizarov, S.; Crespo, J.G. Analysis of solvent flux through nanofiltration membranes by mechanistic, chemometric and hybrid modelling. *J. Membr. Sci.* **2007**, *300*, 191–204. [CrossRef]
28. Hernández, A.; Calvo, J.I.; Prádanos, P.; Tejerina, F. Pore size distributions in microporous membranes. A critical analysis of the bubble point extended method. *J. Membr. Sci.* **1996**, *112*, 1–12. [CrossRef]
29. Yu, J.; Hu, X.; Huang, Y. A modification of the bubble-point method to determine the pore-mouth size distribution of porous materials. *Sep. Purif. Technol.* **2010**, *70*, 314–319. [CrossRef]
30. Esbensen, K.H.; Swarbrick, B. *Multivariate Data Analysis: An Introduction to Multivariate Analysis, Process Analytical Technology and Quality by Design*, 6th ed.; CAMO Software AS: Oslo, Norway, 2018.
31. Erikson, L.; Johansson, E.; Kettaneh-Wold, N.; Trygg, J.; Wikström, C.; Wold, S. *Multi-And Megavariate Data Analysis: Basic Principles and Applications*; Umetrics AB: Umeå, Sweden, 2006.
32. Wold, H. Soft modelling, The basic design and some extensions. In *Systems Under Indirect Observation, Causality-Structure-Prediction*; Part 2; Jöreskog, K.G., Wold, H., Eds.; North-Holland Publishing Co.: Amsterdam, The Netherlands, 1982.
33. Wold, S.; Albano, C.; Dunn, W.; Edlund, U.; Esbensen, K.; Geladi, P.; Hellberg, S.; Johanson, E.; Lindberg, W.; Sjöström, M. Multivariate data analysis in chemistry. In *Mathematics and Statistics in Chemistry*; Kowalski, B.R., Ed.; Reidel Publishing Company: Dordrecht, The Netherlands, 1984.
34. Wold, S.; Johansson, E.; Cocchi, M. PLS-partial least squares projections to latent structures. In *3D QSAR in Drug Design, Theory, Methods, and Applications*; Kubinyi, H., Ed.; ESCOM Science Publishers: Leiden, The Netehrlands, 1993; pp. 523–550.
35. Hotelling, H. Analysis of a complex of statistical variables into principal components. *J. Ed. Psychol.* **1993**, *24*, 417–441. [CrossRef]
36. Fane, A.G.; Fell, C.J.D. A review of fouling and fouling control in ultrafiltration. *Desalination* **1987**, *62*, 117–136. [CrossRef]
37. Lin, C.F.; Lin, A.Y.C.; Chandana, P.S.; Tsai, C.Y. Effects of mass retention of dissolved organic matter and membrane pore size on membrane fouling and flux decline. *Water Res.* **2009**, *43*, 389–394. [CrossRef]
38. Jiraratananon, R.; Chanachai, A. A study of fouling in the ultrafiltration of passion fruit juice. *J. Membr. Sci.* **1996**, *111*, 39–48. [CrossRef]

chemengineering

MDPI

Article

Preliminary Equipment Design for On-Board Hydrogen Production by Steam Reforming in Palladium Membrane Reactors

Marina Holgado and David Alique *

Department of Chemical, Energy and Mechanical Technology, Rey Juan Carlos University, C/Tulipán s/n, 28933 Móstoles, Spain; marinaholgadododones@gmail.com
* Correspondence: david.alique@urjc.es; Tel.: +34-914887603; Fax: +34-914887068

Received: 31 October 2018; Accepted: 7 January 2019; Published: 15 January 2019

Abstract: Hydrogen, as an energy carrier, can take the main role in the transition to a new energy model based on renewable sources. However, its application in the transport sector is limited by its difficult storage and the lack of infrastructure for its distribution. On-board H_2 production is proposed as a possible solution to these problems, especially in the case of considering renewable feedstocks such as bio-ethanol or bio-methane. This work addresses a first approach for analyzing the viability of these alternatives by using Pd-membrane reactors in polymer electrolyte membrane fuel cell (PEM-FC) vehicles. It has been demonstrated that the use of Pd-based membrane reactors enhances hydrogen productivity and provides enough pure hydrogen to feed the PEM-FC requirements in one single step. Both alternatives seem to be feasible, although the methane-based on-board hydrogen production offers some additional advantages. For this case, it is possible to generate 1.82 kmol h^{-1} of pure H_2 to feed the PEM-FC while minimizing the CO_2 emissions to 71 g CO_2/100 km. This value would be under the future emissions limits proposed by the European Union (EU) for year 2020. In this case, the operating conditions of the on-board reformer are T = 650 °C, P_{ret} = 10 bar and H_2O/CH_4 = 2.25, requiring 1 kg of catalyst load and a membrane area of 1.76 m^2.

Keywords: hydrogen; on-board; steam reforming; ethanol; methane; membrane reactor; palladium; modeling

1. Introduction

The current energy model, mainly based on fossil fuels, presents two main drawbacks (i) limitation of reservoirs, which are getting scarcer and, consequently, increasing the price; and (ii) generation of CO_2 emissions during their combustion, definitively contributing to global warming [1]. Under this situation, especially critical in big cities, it is clear there is a need for a new and sustainable energy model, based on renewable energies, capable of facing the increasing energy demand associated with the exponential growth of the global population and, simultaneously helping the reduction of pollutant emissions [2]. This modification of the energy model will not be immediate nor focused on a unique energy source, so a progressive transition in the short-middle term is necessary to achieve a completely stable and safe energy grid while fossil fuels gradually cease [3]. Most perspectives establish hydrogen as the key for this transition due to its high performance and absence of CO_2 emissions during its combustion. It is considered a clean energy carrier because it allows both storage of diverse primary energy sources, renewable ones in an ideal situation, and transformation into different forms of energy, i.e., electrical energy in fuel cells [1]. In this manner, a hydrogen-based energy model could combine at the same time traditional fossil fuels with other renewable sources, while minimizing the environmental impact [4]. However, the lack of highly efficient storage devices and distribution infrastructures is slowing down its real penetration into the system, especially with regard to the transport sector [5,6]. Development of on-board hydrogen production systems would be a great solution to overcome these limitations, generating the H_2 just inside the vehicle from other compounds and, thus, minimizing its difficult storage and transport [7]. However, this application needs to be carefully addressed, especially in terms of dimensions and weight of the on-board H_2 production unit due to the space restrictions in an average vehicle and optimization of power requirements. Considering the reduction of CO_2 emissions as one of the main reasons to use hydrogen in transport, renewable sources need to be pursued for its production, preferably being also easily stored and distributed [8]. Bio-ethanol [9] and bio-methane [10] could accomplish reasonably these requirements. However, it is important to note that the purity of these compounds could affect the associated H_2 production process and storage requirements. In fact, bio-ethanol is usually accompanied by a significant amount of water, while bio-methane could be together with 20%–40% carbon dioxide. Both alternatives have been widely studied due to the use of mature techniques, i.e., steam-reforming [11–14], although only few of them address on-board production for their future application to the transport sector [15]. Among them, only some laboratory-scaled works or modeling studies using small fuel cells of 1–5 kW can be found [15], producing hydrogen via steam reforming, although it is really rare to find the combination of dimensioning the production unit with the polymer electrolyte membrane fuel cells (PEM-FC) requirements in real conditions required for most of typical vehicles (in terms of power and available space). Therefore, the viability of this strategy to power bigger fuel cells, as most of prototypes demand, needs to be properly addressed. Moreover, requirements of H_2 purity are especially important as the PEM-FC (widely proposed for H_2-vehicles) can be easily poisoned with trace amounts of CO [16], and available space inside a typical vehicle for a purification unit is very limited.

In this context, the use of membrane reactors, which combine both chemical reaction and separation steps in a single device, appears as a very attractive alternative for efficient process intensification [17,18]. Selective permeation of hydrogen through an adequate membrane shifts the equilibrium, enhancing the chemical reaction and, thus, improving both conversion and global efficiency while a high-purity product is simultaneously obtained in the permeate side [4,17,18]. Over recent years, multiple experimental and modeling works with membrane reactors can be found in the literature for diverse processes, mainly steam reforming [19], auto-thermal reforming [20], and water gas shift [21]. Most of them present a multi-tubular structure in which the catalyst is placed as a fixed-bed [22] or fluidized-bed [23,24] and the tubes are made of H_2-selective material, usually Pd or Pd-based alloys with high perm-selectivity and good thermal resistance [25]. However, the study of these systems for direct H_2 production on-board is still scarce [15].

In this context, the present work analyzes the use of membrane reactors for ultra-pure H_2 production on-board, capable to power a typical PEM-FC, feeding directly previously purified bio-ethanol or bio-methane in the vehicle. The most convenient operating conditions were studied through modeling to enhance the H_2 production, maximizing the permeation rate and, thus, the chemical reaction displacement, while assuring both thermal and mechanical stability. The reactor design (catalyst load and membrane area) was performed while taking into account main limitations of available space in vehicles. Finally, some considerations about energy integration, economy, and environmental impact were also addressed.

2. Experimental Details

2.1. Process Design for Hydrogen Production On-Board

The process design was focused on the hydrogen requirements of the considered fuel cell, in this case a PEM-FC type. According to the Technology Road Map published for fuel cells and H_2 transition [26], the recommended power energy for utility vehicles goes from 80 to 120 kW with higher heating value efficiencies (HHV) up to 60% (ratio between fuel cell power and high heating value of the gases fed to the anode). Considering the minimum value of this range (80 kW) and typical efficiencies, the on-board production system would need to supply around 1.70 kmol·h^{-1} of pure H$_2$, which was taken as the target value for this work.

Figure 1 presents a block diagram for the entire process designed in the present work. Two main deposits contain the main reactants of the process: fuel (ethanol or methane from previous bio-production processes) and water. Here, it is important to note that a previous purification and conditioning of fuels were considered, feeding the vehicle with pure compounds for easier comparison between both alternatives. Inside the vehicle, the reactants are pre-treated to reach the operating conditions prior to entering the membrane reactor. Basically, this pre-treatment consists of pumping and heating the reactants (vaporizing in case of liquids) until reaching the operating conditions. Then, H$_2$ is produced in the membrane reactor unit (R-1), being simultaneously separated through a palladium membrane to feed the fuel cell (low-temperature polymer electrolyte membrane fuel cell, LT-PEMFC). The H$_2$ flux needs to be cooled and stabilized in a buffer to enlarge the PEM-FC life cycle. The retentate coming from R-1 is fed to a combustor, where the non-converted reactants (bio-ethanol or bio-methane), CO, and non-permeated H$_2$ are burnt to provide the required energy for both pre-treatment units and R-1. Water is separated from CO$_2$ by condensation and then returned to the intake deposit.

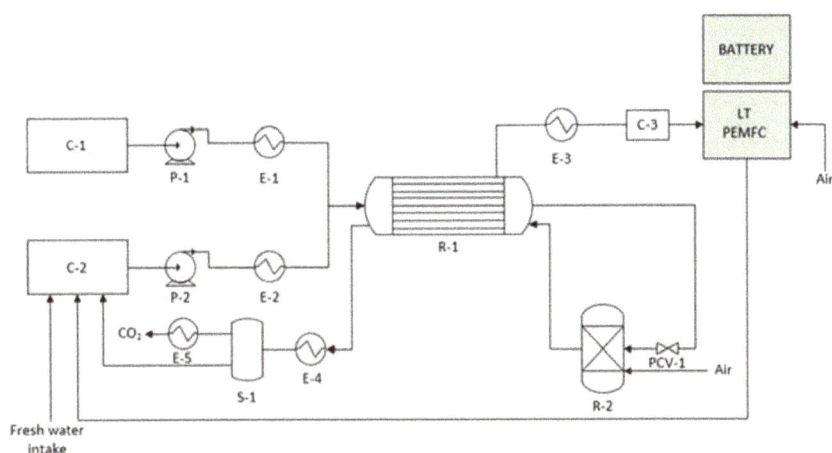

Figure 1. Block diagram for H$_2$-production on-board.

2.2. Membrane Reactor Modeling

The entire process for H$_2$-production on-board, the membrane reactor was modeled in Aspen-Plus® v.10, selecting the SR-POLAR method as the thermodynamic model for the calculations. However, the software does not have a specific block to simulate membrane reactors, in which both the chemical reaction and the product separation are carried out simultaneously. To overcome this problem, the multi-tubular membrane reactor R-1 designed for this work was simulated by successive modules emulating both chemical reaction and H$_2$ separation. Thus, R-1 is divided into a limited number of units formed by a chemical reactor (RPLUG) and a consecutive separator to perform the real membrane reformer. This scheme is usually adopted as a good solution for simulating accurately the ideal shift effect of the membrane in a membrane reactor when only considering the thermodynamic equilibrium [27,28]. In this work, besides the equilibrium displacement, the kinetics of the possible chemical reactions were also considered as detailed in the following section. Figure 2 shows a simplified diagram of the block scheme used to simulate the membrane reactor in Aspen-Plus® v.10.

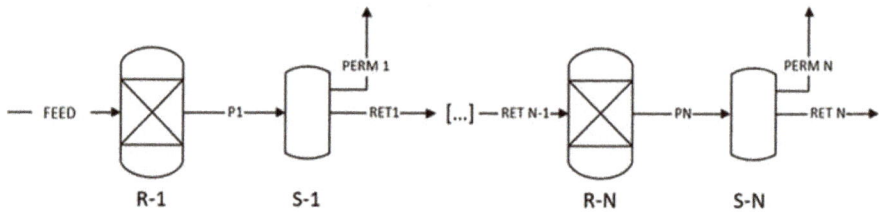

Figure 2. Membrane reactor simulation by finite elements (micro-reactor + micro-separator).

The H_2 permeation through the membrane for each separator unit was calculated following Sievert's Law [29].

$$F_{H_2} = k_{H_2} \times A \times \left(P_{H_2,ret}^{0,5} - P_{H_2,perm}^{0,5} \right) \quad (1)$$

where F_{H_2} represents the hydrogen permeate flow in mol/s, k_{H_2} the H_2 permeance, A is the permeation area, $P_{H_2,ret}$ and $P_{H_2,perm}$ the hydrogen partial pressure in retentate and permeate sides, respectively. For this study, a general permeance of $k_{H_2} = 2.43 \times 10^{-3}$ mol·m^{-2}·s^{-1}·Pa$^{-0.5}$ was considered, taking as reference the DOE (Department of Energy of United States of America) technical targets for dense metallic membranes, in which 300 scfh/ft^{-2} hydrogen flow-rate is recommended when operating under 150 and 50 psia hydrogen partial pressure in retentate and permeate sides, respectively [30]. The permeate side was maintained at ambient pressure (1 bar) without applying any gas carrier, while the partial pressure of hydrogen in the retentate side was calculated by multiplying the operating pressure of the reactor and the hydrogen fraction present in the products that leave the previous reactor unit. This assumption is very realistic if considering a low pressure drop inside the reactor due to its considered length and the common control of pressure with back-regulators. The membrane area used in the Sievert's Law for each separator unit will be the total membrane area considered in the study divided by the number of separator units that emulate the membrane reactor. Once the permeated hydrogen has been calculated, the split fraction is obtained by dividing this value by the total hydrogen that has entered this separator unit. No sweep gas was considered in the permeate stream in order to obtain ultra-pure hydrogen that feeds the fuel cell and, consequently, powers the vehicle.

2.3. Chemical Reactions and Kinetics

As it was previously mentioned, not only the chemical equilibrium was considered but the kinetics for diverse possible reactions inside the membrane reactors were also taken into account for the modeling. Chemical reactions considered in the present work were collected from diverse experiments from literature and the most relevant ones can be summarized as follows:

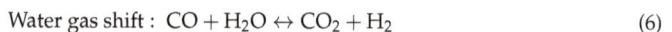

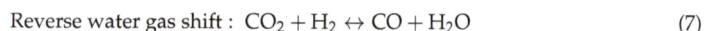

$$\text{Ethanol decomposition}: C_2H_5OH \rightarrow CO + CH_4 + H_2 \quad (2)$$

$$\text{Ethanol steam reforming}: C_2H_5OH + H_2O \rightarrow CO_2 + CH_4 + 2H_2 \quad (3)$$

$$\text{Methane steam reforming (i)}: CH_4 + 2H_2O \leftrightarrow CO_2 + 4H_2 \quad (4)$$

$$\text{Methane steam reforming (ii)}: CH_4 + H_2O \leftrightarrow CO + 3H_2 \quad (5)$$

$$\text{Water gas shift}: CO + H_2O \leftrightarrow CO_2 + H_2 \quad (6)$$

$$\text{Reverse water gas shift}: CO_2 + H_2 \leftrightarrow CO + H_2O \quad (7)$$

Based on previous publications from Llera et al. [31] and Hou et al. [32] for ethanol and methane steam reforming, respectively, Langmuir-Hinshelwood (LHHW) kinetics have been implemented in this work for modeling all described possible chemical reactions. LHHW equations involve each adsorption, reaction, and desorption steps carried out during the chemical reaction, thus providing more precise results than a Power Law model. All details about the kinetics expressions used in

Aspen-Plus® v.10 are included in Appendix A. Both reactor dimensions and, therefore, the residence time, vary during the modeling for the optimization of the membrane area in the present study.

3. Results and Discussion

3.1. Preliminary Membrane Reactor Design: Modeling and Main Operating Conditions

An adequate membrane reactor design involves the selection of specific catalysts for principal chemical reactions, catalyst load, membrane characteristics, device dimensioning (including the required permeation area), and main operating conditions. As it was previously stated, proper catalysts for each alternative addressed in the present work, as well as the associate kinetics parameters, were taken from literature [31,32]. Thus, the present section is focused on modeling the membrane reactor and analyzing the main operating conditions. The adopted strategy for modeling the membrane reactor is based on dividing the equipment in a limited number of consecutive RPLUG reactor and separator blocks. Thus, the first task was to determine the optimal number of units for simulating the shift effect of the reaction thanks to the simultaneous H_2 permeation through the membrane. This study was carried out with some preliminary operating conditions, including a total catalyst load of 35 kg and a maximum possible membrane area (2.42 m^2) that fits in the available space inside the vehicle. This area was divided into equal parts for each considered number of simulation units. Depicted in Figure 3, we found both the permeate flux and the total amount of H_2 generated from ethanol (Figure 3a) or methane (Figure 3b) in the membrane reactor for an increasing number of simulation units (reactor-separator). In general, H_2 production increases as a greater number of simulation units is considered due to the shift effect on the thermodynamic equilibrium; while a contrary effect can be observed on permeate flux, due to the H_2 depletion along the axial dimension of the reactor. For both ethanol and methane intakes, H_2 production seems to stabilize after 10 simulation units, so this value was selected to continue the study and analyze in detail the best operating conditions.

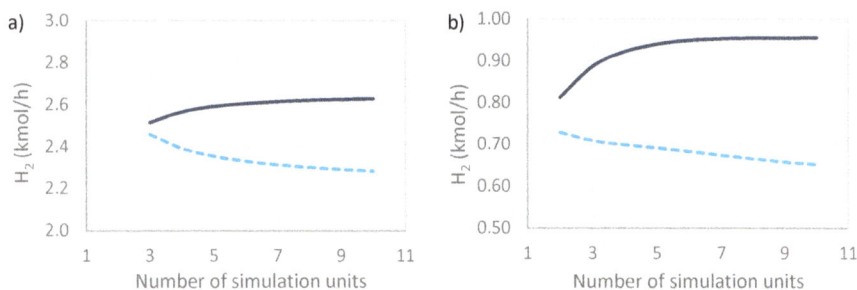

Figure 3. Determination of simulation units for the membrane reactor when feeding the process with: (**a**) ethanol (feed = 10 kmol/h, T = 600 °C, P = 10 bar, H_2O/feed = 2) and (**b**) methane (feed = 10 kmol/h, T = 600 °C, P = 10 bar, H_2O/feed = 2). Legend: continuous line = total generated H_2 and dashed line = permeated H_2.

After determining the suitable number of simulation units for the membrane reactor, the influence of temperature, H_2O/feed ratio and pressure were addressed as the main operating parameters for both ethanol and methane feeding the membrane reactor (Figure 4). For these studies, 35 kg catalyst load and 2.42 m^2 of membrane area were maintained, keeping in mind that they were provisional values to be optimized afterwards.

The temperature optimization was first studied maintaining a pressure of 10 bar and the steam to feed ratio at four and three in the case of reforming ethanol or methane, respectively. As it can be observed in Figure 4a, an increasing temperature favors the hydrocarbons transformation into hydrogen. The main ethanol decomposition (Equation (2)) and steam reforming (Equations (3)–(5)) are endothermic reactions so they will be thermodynamically improved by increasing temperature.

Additionally, both reaction kinetics and hydrogen permeance through the membrane are also increased with temperature, following an Arrhenius-type dependence, so the shift effect of the membrane reactor is also boosted and, consequently, the hydrogen production rate. Thus, it can be stated that, in general, higher temperatures improve the hydrogen production. However, temperature is limited by the thermal stability of the H_2-selective membrane. Pd-based membranes are prepared onto supporting materials and experimentally they are used in the typical range of 400–550 °C to prevent possible damages on the composite structure, although it is expected to resist slightly higher temperatures [33,34]. In this manner, it is also possible to find several works in which these membranes operate at temperatures up to 650 °C with satisfactory results in terms of mechanical stability [35–38]. Under this perspective, and considering that temperatures above 650 °C do not increase the hydrogen productivity remarkably, this value was selected as the most appropriate operating temperature to perform the process when feeding both ethanol and methane.

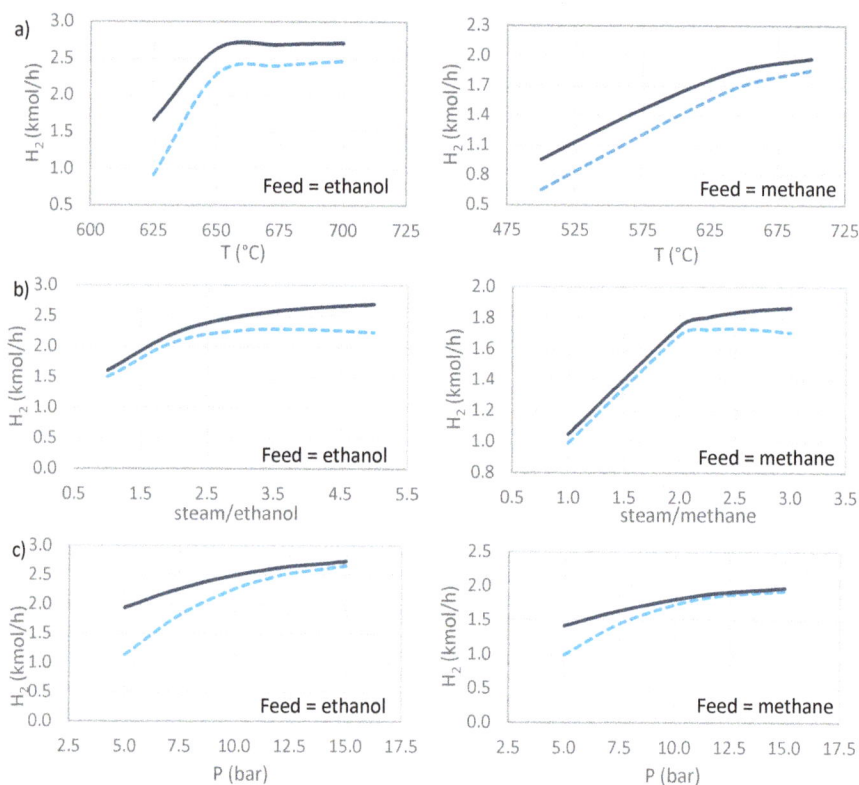

Figure 4. Influence of main operating conditions for the membrane reactor when feeding ethanol or methane: (**a**) temperature, (**b**) steam to feed hydrocarbon ratio and (**c**) total retentate pressure. Legend: continuous line = total generated H_2 and dashed line = permeated H_2.

Steam to feed hydrocarbon ratio in the membrane reactor was the next operating condition analyzed in this work (Figure 4b). For this study, the reactor temperature was maintained at 650 °C, considering the optimum value obtained in the previous study, and the reactor pressure at a preliminary value of 10 bar. As it can be extracted from the results, total hydrogen production is enhanced by increasing values of steam content in the feed. In general, the presence of water promotes all chemical reactions in which it acts as reactant, shifting the thermodynamic equilibrium towards further hydrogen production, according to Le Châtelier's principle. However, large quantities of water dilute the generated hydrogen, reducing its partial pressure in the retentate side and, in consequence, the driving force of the permeation process. Thus, the thermodynamic equilibrium shift due to the H_2 extraction through the membrane is limited. These opposite effects can explain the results collated in Figure 4b. First, the permeate flow-rate increases as steam to feed hydrocarbon ratio increases, as the chemical reaction improvement is more important than the hydrogen dilution effect. A maximum value around the ratio 2–3 is reached for the permeate flow-rate, the dilution effect becoming greater than the chemical reaction improvement from this point. This effect is caused by a drastic hydrogen partial pressure decrease in the retentate side, also affecting the pure hydrogen recovery. Considering this behavior and the energy requirements for heating the entire feed stream to the membrane reactor (including both hydrocarbon and steam), values of 3.00 and 2.25 were selected for steam-to-ethanol and steam-to-methane ratio, respectively.

Finally, the operating pressure in the membrane reactor was also evaluated, taking the optimal values obtained in previous studies for the temperature and steam to feed ratio. The modeling performance at these conditions is shown in Figure 4c. In general, an increase in both total produced H_2 and permeate H_2 can be observed as the pressure increases in the retentate side, being able to extract almost all the H_2 generated in the membrane reactor as a pure gas in the permeate side at pressures higher than 10 bar. This separation is slightly easier in the case of feeding methane instead of ethanol due to the relationship between membrane area (kept constant) and the total H_2 generated. A pressure increase clearly makes the H_2 separation through the membrane easier due to the associated improvement of the permeation driving force. The higher the H_2 permeation, the greater the shift effect on the thermodynamic equilibrium for chemical reactions R1 to R4, therefore enhancing the global productivity. However, in the case of ethanol steam reforming, previous studies found a decrease of H_2 yield in a traditional fixed-bed reactor due to the production of methane, which cannot be easily converted into hydrogen at high pressures as the reaction is shifted towards the reactants, accordingly to Le Châtelier's principle [39]. Of course, this negative effect on the H_2 production is clear in the case of directly feeding methane to the process. However, as it can be extracted from the modeling results, this negative effect on the thermodynamics seems to be compensated by the continuous hydrogen removal through the membrane, obtaining a steady increase in hydrogen production as the pressure rises in the studied range. Considering other aspects related to the mechanical stability of the membrane, a total pressure of 10 bar for the retentate side was selected as the most appropriate operating condition to be used in the membrane reactor, both for ethanol and methane intakes. The stability of H_2-selective membranes at this operating pressure was demonstrated in previous works at lab scale, avoiding cracks generation or delamination of the selective film [36,40].

3.2. Reactor Design Optimization

After analyzing the main operating parameters for the membrane reactor, the equipment design was optimized in terms of catalyst load, required permeation area, possible energy integration, and some preliminary economic considerations, aiming to produce 1.70 kmol·h^{-1} of pure H_2 to power the considered 80 kW PEM-FC for the vehicle.

The maximum size considered for the membrane reactor is based on the available space under the rear seats of a standard vehicle, where it will be placed. In order to maximize the compactness of the reformer and maximize the permeation area, a multi-tubular reactor configuration was adopted. In this reactor, a triangular distribution of membranes with 1.25 in of inter-tubular space to ensure good heat transfer was considered, thus being possible to accommodate a maximum of 38 tubes of 80 cm in length. This reached a maximum membrane area of 2.42 m^2.

Theoretically, the amount of catalyst affects the space velocity inside the reactor, the hydrocarbon conversion and, consequently, the amount of produced hydrogen. Figure 5 collates the results obtained for different catalyst loads when feeding the membrane reactor with ethanol (Figure 5a) and methane (Figure 5b). A stable H_2 production rate is reached for a catalyst load of 5 kg and 1 kg when feeding the reactor with ethanol and methane, respectively. The facility to reach a stable value with a smaller amount of catalyst in the second case can be explained by the kinetic parameters. The pre-exponential factors for the kinetic expression of methane steam reforming (Equation (4)) are higher by several orders of magnitude than the corresponding ones for ethanol steam reforming (Equation (3)). It means that higher conversions can be achieved with a smaller catalyst load. In the case of feeding the membrane reactor with ethanol, almost complete conversions (around 99.99%) were achieved for catalyst loads greater than 5 kg, while this value was maintained around 40% in the case of using 1 kg of catalyst.

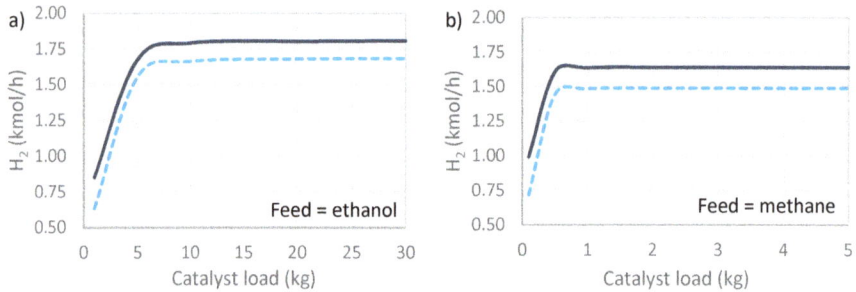

Figure 5. Influence of catalyst load when feeding: (**a**) ethanol and (**b**) methane. Legend: continuous line = total generated H_2 and dashed line = permeated H_2.

Then, several combinations of catalyst load, membrane area and reactant feed were tested aiming to achieve the production target of 1.70 $kmol·h^{-1}$ of pure H_2. With these simulations we can see that, despite hydrogen production being maintained as very stable for increasing catalyst loads, the associated decrease of space velocity improves the hydrogen recovery, as it is possible to save part of the initially considered membrane area. In this manner, it could be possible to maintain similar hydrogen production, saving 20% of membrane area by doubling the catalyst load from 5 kg to 10 kg. However, by doubling again the catalyst load from 10 kg to 20 kg this effect would only save another 3% of the membrane area. Thus, 10 kg was chosen as the optimum catalyst load for the ethanol steam reformer. This effect is negligible in the case of considering a methane feed.

The optimal combination was found to be 1.87 m^2 membrane area, 10 kg of catalyst load, and 0.37 $kmol\,h^{-1}$ feed for the ethanol steam reforming and 1.76 m^2, 1 kg of catalyst and 0.54 $kmol·h^{-1}$ when feeding methane. As it was previously described, these membrane areas were achieved by considering the use of a multi-tubular membrane reactor. The recent trends in membrane preparation used for hydrogen production processes have been directed to synthesize composite membranes in which a thin layer of palladium or a palladium-based alloy is deposited onto a porous supporting material [34]. Selecting standard dimensions for these supports, i.e., outside diameter of 1.0 in and total length around 24 in, the membrane reactor design will require 28 membranes, ensuring a good fit to the available space under the rear seats in any utility vehicle.

The energy requirements for the proposed model, including the reactants pre-treatment and the heat of reaction requirements, can be achieved by combusting the retentate gases from the membrane reactor. Thus, both ethanol/methane intake and membrane area were optimized to reach an autonomous process in terms of energy, while the desired pure-H_2 to feed the PEM-FC was maintained. The transitory state until achieving this situation was not taken into account in the present study. Before the reactor reaches the optimum operating temperature to act as an autonomous system, ethanol or methane would need to be fed directly to the combustor, providing the necessary energy for the initial process conditioning. Thus, the total fuel consumption would be slightly higher than that indicated in this work.

On analyzing the convenience of using a membrane reactor for the on-board H_2 generation instead of a traditional reaction system, clear benefits can be found. In this manner, a 75% rise in H_2 production is achieved by using a membrane reactor for the on-board hydrogen production from ethanol compared to that obtained in a conventional reactor operating under the same conditions (T = 650 °C, P_{ret} = 10 bar, H_2O/C_2H_5OH = 3 and m_{cat} = 10 kg). Thus, the H_2 flow-rate is increased from 1.06 kmol h^{-1} to 1.85 kmol·h^{-1} due to the total conversion of the ethanol being reached, whereas it was maintained below 95% in a conventional reactor. In the case of feeding the system with methane, a similar behavior can be found, although in this case the increase was greater. Hydrogen productivity and methane conversion were increased from 0.9 kmol·h^{-1} to 1.82 kmol·h^{-1} and from 42.8% to 84.9%, respectively, operating under the same conditions for both conventional and membrane reactors (T = 650 °C, P_{ret} = 10 bar, H_2O/CH_4 = 2.25 and m_{cat} = 1 kg). All these parameters are summarized in Table 1. The results also imply additional benefits in both weight and space savings due to the intensification of the process reached with the membrane reactor.

Table 1. Optimization results for the membrane reactor design.

Feed HC	N_{in} (kmol h^{-1})	$A_{membrane}$ (m^2)	Catalyst Load (kg)	T (°C)	$P_{retentate}$ (bar)	H_2O/HC	$N_{H2, out}$ (kmol h^{-1})	X_{HC} (%)	Net Energy Balance (kW)
C_2H_5OH	0.37	1.87	10	650	10.0	3.00	1.85	100	−3.13
CH_4	0.54	1.76	1	650	10.0	2.25	1.82	84.9	−2.99

Finally, a preliminary approach to main economic and environmental aspects was also included to analyze roughly the viability of on-board H_2 production via membrane reactors for powering vehicles. First commercial hydrogen vehicles on the market have an autonomy of around 500–650 km with a pressurized hydrogen tank of 5 kg [41,42]. Considering conservative criteria, a preliminary consumption of 1 kg H_2/100 km for general PEM-FC vehicles is estimated. If hydrogen on-board generation is assumed, there is no need to store it, thus solving one of the main drawbacks for commercially introducing hydrogen vehicles in the near future. Considering a regular deposit of 55 L for liquid fuels (i.e., ethanol coming from bio-routes), and simulated results of this work, 8.5 kg of hydrogen could be generated, giving an autonomy for the car of around 850 km. On the other hand, considering a 15 kg deposit for the methane alternative (value used in current natural gas-powered vehicles in the market), 5.8 kg of hydrogen could be produced, making it possible to cover around 580 km. The membrane would be one of the most expensive elements in the process; estimating its cost from economic targets proposed by the US Department of Energy for ensuring commercial viability of this technology, assuming a cost of 500 US $/ft^2, it would mean, around 1520 € m^2 with the current exchange rate [30]. In this context, the cost for the membranes of the on-board reformers would be around 2650–2800 € in the case of considering feeding the vehicle with bio-methane or bio-ethanol, respectively. This cost can be easily absorbed by both manufacturers and customers, especially considering that CO_2 emissions can be significantly reduced. The European Commission indicates that CO_2 emissions need to be maintained below 95 g·km^{-1} for year 2020 [43]. Under this perspective, the studied process feeding with bio-ethanol presents potential CO_2 emissions of around 97 g CO_2/100 km, slightly higher than the limit proposed by the UE. However, feeding the system with methane, this value could be reduced to 71 g CO_2/100 km due to its higher H/C ratio. Here, it is important to remember that previous purification processes for bio-ethanol and bio-methane feedstock have been considered, feeding the vehicle with the pure compounds to avoid a reduction of the useful volume of the vehicle tank. Both of them represent good results, noticeably reducing current values for latest gasoline and diesel vehicles, with CO_2 emission levels of 123 and 119 g CO_2/100 km, respectively.

4. Conclusions

This work addressed a first approach for analyzing the viability of H_2 on-board production by membrane reactors in PEM-FC vehicles via mathematical modelling with Aspen-Plus$^®$ v.10. Despite further experimental studies needing to be performed, some interesting insights can be extracted for alternative hydrogen production from bio-ethanol or bio-methane. Firstly, it was demonstrated that the use of membrane reactors enhances the hydrogen productivity and provides enough pure hydrogen to feed the PEM-FC requirements in one single step. Operating conditions for both alternatives were optimized, studying the effect of temperature, pressure, steam/hydrocarbon ratio, and catalyst load for each case. The methane-based on-board hydrogen production seems to be the best alternative, generating 1.82 kmol·h^{-1} of pure H$_2$ for feeding the PEM-FC and minimizing the CO$_2$ emissions up to 71 g CO$_2$/100 km, ensuring the future limitation proposed by the UE for year 2020 is achieved. This alternative ensures an autonomy of around 580 km for the H$_2$-vehicle assuming a typical methane deposit of similar capacity to the gas-feed commercially available vehicles. For this alternative, the on-board reformer operates at T = 650 °C, P_{ret} = 10 bar, and H$_2$O/CH$_4$ = 2.25, requiring 1 kg of catalyst load and a membrane area of 1.76 m^2. However, the alternative achieved from bio-ethanol can also be considered for the future, reaching similar results (1.87 kmol·h^{-1} of pure H$_2$ at comparable operating conditions) and ensuring a more realistic production from renewable routes in the required terms.

Author Contributions: D.A. and M.H. conceived and designed the experiments; M.H. performed the experiments; D.A. and M.H. analyzed the data; no reagents nor materials were necessary for this study; M.H. wrote the paper with revision of D.A.

Funding: This research received no external funding apart the facilities of Rey Juan Carlos University for studying Chemical Engineering and the collaboration scholarship for M.H. in the Department of Chemical, Energy, and Mechanical Technology of the above-mentioned university.

Acknowledgments: The authors of this work are hugely grateful for the support achieved from Rey Juan Carlos University (Spain). M. Holgado especially acknowledges the collaboration scholarship of 8 months in the Department of Chemical, Energy, and Mechanical Technology of the above-mentioned university. Additionally, we also thanks prof. José Antonio Calles for his advises in preliminary works of this research.

Conflicts of Interest: The authors declare no conflict of interest.

Appendix A

The LHHW kinetics expressions considered in the present work for the ethanol steam reforming are summarized as follows:

$$r_{R1} = \frac{k_{R_1} y_{Et} y_{CH_4}^{-1} y_{H_2}^{-1/2}}{DEN^2} \tag{A1}$$

$$r_{R2} = \frac{k_{R_2} y_{Et} y_{H_2O} y_{CH_4}^{-1} y_{H_2}^{-1}}{DEN^2} \tag{A2}$$

$$r_{R3} = \frac{k_{R_3} y_{H_2O}^2 y_{CH_4} y_{H_2}^{-5/2}(1 - \beta_{R3})}{DEN^3} \tag{A3}$$

$$r_{R6} = \frac{k_{R_6} y_{CO_2} y_{H_2}^{1/2}(1 - \beta_{R6})}{DEN^2} \tag{A4}$$

$$
\begin{aligned}
DEN = {} & 1 + K_{Et} y_{Et} + K_{Et} y_{Et} y_{H_2}^{-1/2} + K_{Ac} y_{Et} y_{H_2}^{-1} + K_{CHO} y_{Et} y_{CH_4}^{-1} y_{H_2}^{-1/2} \\
& + K_{CH_3} y_{CH_4} y_{H_2}^{-1/2} + K_{CH_2} y_{CH_4} y_{H_2}^{-1} + + K_{CH} y_{CH_4} y_{H_2}^{-3/2} \\
& + K_{H_2O} y_{H_2O} + K_{OH} \, y_{H_2O} y_{H_2}^{-1/2} + K_{CH_4} y_{CH_4} + K_{CO} y_{CO} \\
& + K_{CO_2} y_{CO_2} + K_H y_{H_2}^{1/2} + K_{H_2} \, y_{H_2}
\end{aligned}
\tag{A5}
$$

where β is the ratio between the product of each component fraction to the stoichiometric coefficient and the equilibrium constant of the considered reaction:

$$\beta_{R3} = \frac{y_{H_2}^4 y_{CO_2}}{K_{R3}^{eq} y_{CH_4} y_{H_2O}^2} \tag{A6}$$

$$\beta_{R6} = \frac{y_{H_2O}y_{CO}}{K_{R6}^{eq}y_{CO_2}y_{H_2}} \tag{A7}$$

At the same time, the equilibrium constant can be defined as the ratio between the kinetics constants for both direct and reverse reactions, obtaining:

$$r_{R3} = \frac{k_{R3}y_{H_2O}^2y_{CH_4}y_{H_2}^{-\frac{5}{2}}\left(1 - \frac{k_{R-3}y_{H_2}^4y_{CO_2}}{k_{R3}y_{CH_4}y_{H_2O}^2}\right)}{DEN^3} = \frac{k_{R3}y_{H_2O}^2y_{CH_4}y_{H_2}^{-5/2}}{DEN^3} - \frac{k_{R-3}y_{CO_2}y_{H_2}^{3/2}}{DEN^3} \tag{A8}$$

$$
\begin{aligned}
r_{R6} &= \frac{k_{R6}y_{CO_2}y_{H_2}^{1/2}\left(1 - \frac{k_{R-6}y_{H_2O}y_{CO}}{k_{R6}y_{CO_2}y_{H_2}}\right)}{DEN^2}\\
&= \frac{k_{R6}y_{CO_2}y_{H_2}^{1/2}}{DEN^2} - \frac{k_{R-6}y_{CO}y_{H_2O}y_{H_2}^{-1/2}}{DEN^2}
\end{aligned}
\tag{A9}
$$

In the case of feeding methane, LHHW kinetics expressions can be described as follows:

$$r_3 = \frac{k_3\left(\frac{P_{CH_4}P_{H_2O}}{P_{H_2}^{1.75}}\right)\left[1 - \left(\frac{P_{CO_2}P_{H_2}^4}{K_{P3}}P_{CH_4}P_{H_2O}^2\right)\right]}{DEN^2} \tag{A10}$$

$$r_4 = \frac{k_4\left(\frac{P_{CH_4}P_{H_2O}^{0.5}}{P_{H_2}^{1.25}}\right)\left[1 - \left(\frac{P_{CO}P_{H_2}^3}{K_{P4}}P_{CH_4}P_{H_2O}\right)\right]}{DEN^2} \tag{A11}$$

$$r_5 = \frac{k_5\left(\frac{P_{CO}P_{H_2O}^{0.5}}{P_{H_2}^{0.5}}\right)\left[1 - \left(\frac{P_{CO_2}P_{H_2}}{K_{P5}}P_{CO}P_{H_2O}\right)\right]}{DEN^2} \tag{A12}$$

$$DEN = 1 + K_{CO}P_{CO} + K_H P_H^{0.5} + K_{H_2O}P_{H_2O}P_{H_2}^{-1} \tag{A13}$$

References

1. Balat, H.; Kırtay, E. Hydrogen from biomass—Present scenario and future prospects. *Int. J. Hydrogen Energy* **2010**, *35*, 7416–7426. [CrossRef]
2. International Energy Agency. *World Energy Outlook 2017*; International Energy Agency: Paris, France, 2018.
3. Crespo, P.; Van Nieuwkoop, R.H.; Kardakos, E.G.; Schaffner, C. Modelling the energy transition: A nexus of energy system and economic models. *Energy Strateg. Rev.* **2018**, *20*, 229–235. [CrossRef]
4. Lu, G.Q.; da Costa, J.C.D.; Duke, M.; Giessler, S.; Socolow, R.; Williams, R.H.; Kreutz, T. Inorganic membranes for hydrogen production and purification: A critical review and perspective. *J. Colloid Interface Sci.* **2007**, *314*, 589–603. [CrossRef] [PubMed]
5. Durbin, D.J. Review of hydrogen storage techniques for on board vehicle applications. *Int. J. Hydrogen Energy* **2013**, *38*, 14595–14617. [CrossRef]
6. Agnolucci, P. Hydrogen infrastructure for the transport sector. *Int. J. Hydrogen Energy* **2007**, *32*, 3526–3544. [CrossRef]
7. Brunetti, A.; Barbieri, G.; Drioli, E. Integrated membrane system for pure hydrogen production: A Pd–Ag membrane reactor and a PEMFC. *Fuel Process. Technol.* **2011**, *92*, 166–174. [CrossRef]
8. Pfeifer, A.; Dobravec, V.; Pavlinek, L.; Kraja, G. Integration of renewable energy and demand response technologies in interconnected energy systems. *Energy* **2018**, *161*, 447–455. [CrossRef]
9. Benito, M.; Sanz, J.L.; Isabel, R.; Padilla, R.; Arjona, R.; Daza, L. Bio-ethanol steam reforming: Insights on the mechanism for hydrogen production. *J. Power Sources* **2005**, *151*, 11–17. [CrossRef]
10. Oreggioni, D.; Reilly, M.; Kirby, E.; Ferrão, P.; Fournier, J.; Corre, L. ScienceDirect ScienceDirect 1st International Conference on Sustainable Energy and Resource Use in Food Chains Techno-economic analysis of bio-methane production on District Heating and Cooling from agriculture and food industry waste Assessing the feasibility of using the heat demand-outdoor temperature function for a long-term district heat demand forecast. *Energy Procedia* **2017**, *123*, 81–88. [CrossRef]

11. Vita, A.; Pino, L.; Italiano, C.; Palella, A. Chapter 6—Steam Reforming, Partial Oxidation, and Autothermal Reforming of Ethanol for Hydrogen Production in Conventional Reactors. In *Ethanol*; Basile, A., Iulianelli, A., Dalena, F., Veziroğlu, T.N., Eds.; Elsevier: New York, NY, USA, 2019; pp. 159–191.
12. Murmura, M.A. Modeling Fixed Bed Membrane Reactors for Hydrogen Production through Steam Reforming Reactions: A Critical Analysis. *Membranes* **2018**, *8*, 34. [CrossRef]
13. Simakov, D.S.A.; Sheintuch, M. Demonstration of a scaled-down autothermal membrane methane reformer for hydrogen generation. *Int. J. Hydrogen Energy* **2009**, *34*, 8866–8876. [CrossRef]
14. Tartakovsky, L.; Sheintuch, M. Fuel reforming in internal combustion engines. *Prog. Energy Combust. Sci.* **2018**, *67*, 88–114. [CrossRef]
15. Purnima, P.; Jayanti, S. ScienceDirect A high-efficiency, auto-thermal system for on-board hydrogen production for low temperature PEM fuel cells using dual reforming of ethanol. *Int. J. Hydrogen Energy* **2016**, *41*, 13800–13810. [CrossRef]
16. Jaggi, V.; Jayanti, S. A conceptual model of a high-efficiency, stand-alone power unit based on a fuel cell stack with an integrated auto-thermal ethanol reformer. *Appl. Energy* **2013**, *110*, 295–303. [CrossRef]
17. Sanz, R.; Calles, J.A.; Alique, D.; Furones, L.; Ordóñez, S.; Marín, P. Hydrogen production in a Pore-Plated Pd-membrane reactor: Experimental analysis and model validation for the Water Gas Shift reaction. *Int. J. Hydrogen Energy* **2015**, *40*, 3472–3484. [CrossRef]
18. Kikuchi, E. Membrane reactor application to hydrogen production. *Catal. Today* **2000**, *56*, 97–101. [CrossRef]
19. De Nooijer, N.; Gallucci, F.; Pellizzari, E.; Melendez, J.; Alfredo, D.; Tanaka, P.; Manzolini, G.; Van Sint, M. On concentration polarisation in a fluidized bed membrane reactor for biogas steam reforming: Modelling and experimental validation. *Chem. Eng. J.* **2018**, *348*, 232–243. [CrossRef]
20. Spallina, V.; Matturro, G.; Ruocco, C.; Meloni, E.; Palma, V.; Fernandez, E.; Gallucci, F. Direct route from ethanol to pure hydrogen through autothermal reforming in a membrane reactor: Experimental demonstration, reactor modelling and design. *Energy* **2018**, *143*, 666–681. [CrossRef]
21. Sanz, R.; Calles, J.A.; Alique, D.; Furones, L. H₂ production via water gas shift in a composite Pd membrane reactor prepared by the pore-plating method. *Int. J. Hydrogen Energy* **2014**, *39*, 4739–4748. [CrossRef]
22. Anzelmo, B.; Liguori, S.; Mardilovich, I.; Iulianelli, A.; Ma, Y. ScienceDirect Fabrication & performance study of a palladium on alumina supported membrane reactor: Natural gas steam reforming, a case study. *Int. J. Hydrogen Energy* **2018**, *43*, 7713–7721. [CrossRef]
23. Ruocco, C.; Meloni, E.; Palma, V.; Annaland, M.V.; Spallina, V.; Gallucci, F. Pt–Ni based catalyst for ethanol reforming in a fluidized bed membrane reactor. *Int. J. Hydrogen Energy* **2016**, *41*, 20122–20136. [CrossRef]
24. Arratibel, A.; Medrano, J.A.; Melendez, J.; Tanaka, D.A.P.; van Sint Annaland, M.; Gallucci, F. Attrition-resistant membranes for fluidized-bed membrane reactors: Double-skin membranes. *J. Membr. Sci.* **2018**, *563*, 419–426. [CrossRef]
25. Plazaola, A.A.; Tanaka, D.A.P.; Annaland, M.V.S.; Gallucci, F. Recent advances in Pd-based membranes for membrane reactors. *Molecules* **2017**, *22*, 51. [CrossRef] [PubMed]
26. International Energy Agency. *Technology Roadmap, Hydrogen and Fuel Cells*; International Energy Agency: Paris, France, 2015.
27. Rocha, C.; Soria, M.A.; Madeira, L.M. Steam reforming of olive oil mill wastewater with in situ hydrogen and carbon dioxide separation—Thermodynamic analysis. *Fuel* **2017**, *207*, 449–460. [CrossRef]
28. Yonamine, W.; Thangavel, S.; Ohashi, H.; Fushimi, C. Performance analysis of a water–gas shift membrane reactor for integrated coal gasification combined cycle plant. *Energy Convers. Manag.* **2018**, *174*, 552–564. [CrossRef]
29. Mejdell, A.L.; Jøndahl, M.; Peters, T.A.; Bredesen, R.; Venvik, H.J. Effects of CO and CO₂ on hydrogen permeation through a ∼3 μm Pd/Ag 23 wt.% membrane employed in a microchannel membrane configuration. *Sep. Purif. Technol.* **2009**, *68*, 178–184. [CrossRef]
30. N.E.T.L. (NETL), Department of Energy (US). *Test Protocol, Testing of Hydrogen Separation Membranes*; 2008. Available online: https://www.netl.doe.gov/.../Membrane-test-protocol-v10_2008_final10092008.pdf (accessed on 26 December 2017).
31. Llera, I.; Mas, V.; Bergamini, M.L.; Laborde, M.; Amadeo, N. Bio-ethanol steam reforming on Ni based catalyst. Kinetic study. *Chem. Eng. Sci.* **2012**, *71*, 356–366. [CrossRef]
32. Hou, K.; Hughes, R. The kinetics of methane steam reforming over a Ni/α-Al₂O catalyst. *Chem. Eng. J.* **2001**, *82*, 311–328. [CrossRef]

33. Alique, D.; Imperatore, M.; Sanz, R.; Calles, J.A.; Baschetti, M.G. Hydrogen permeation in composite Pd-membranes prepared by conventional electroless plating and electroless pore-plating alternatives over ceramic and metallic supports. *Int. J. Hydrogen Energy.* **2016**, *41*, 19430–19438. [CrossRef]

34. Alique, D. Processing and Characterization of Coating and Thin Film Materials. In *Advanced Ceramic and Metallic Coating and Thin Film Materials for Energy and Environmental Applications*; Zhang, J., Jung, Y., Eds.; Springer: Cham, Switzerland, 2018. [CrossRef]

35. El Hawa, H.W.A.; Paglieri, S.N.; Morris, C.C.; Harale, A.; Way, J.D. Identification of thermally stable Pd-alloy composite membranes for high temperature applications. *J. Membr. Sci.* **2014**, *466*, 151–160. [CrossRef]

36. Hedayati, A.; Le, O.; Lacarri, B. Dynamic simulation of pure hydrogen production via ethanol steam reforming in a catalytic membrane reactor. *Energy* **2016**, *117*, 316–324. [CrossRef]

37. Hedayati, A.; Le, O.; Lacarrière, B.; Llorca, J. Experimental and exergy evaluation of ethanol catalytic steam reforming in a membrane reactor. *Catal. Today* **2016**, *268*, 68–78. [CrossRef]

38. Jia, H.; Wu, P.; Zeng, G.; Salas-Colera, E.; Serrano, A.; Castro, G.R.; Xu, H.; Sun, C.; Goldbach, A. High-temperature stability of Pd alloy membranes containing Cu and Au. *J. Membr. Sci.* **2017**, *544*, 151–160. [CrossRef]

39. Ebshish, A.; Yaakob, Z.; Narayanan, B.; Bshish, A. Steam Reforming of Glycerol over Ni Supported Alumina Xerogel for Hydrogen Production. *Energy Procedia* **2012**, *18*, 552–559. [CrossRef]

40. Barreiro, M.M.; Maroño, M.; Sánchez, J.M. Hydrogen permeation through a Pd-based membrane and RWGS conversion in H_2/CO_2, $H_2/N_2/CO_2$ and $H_2/H_2O/CO_2$ mixtures. *Int. J. Hydrogen Energy* **2014**, *39*, 4710–4716. [CrossRef]

41. Lipman, T.E.; Elke, M.; Lidicker, J. ScienceDirect Hydrogen fuel cell electric vehicle performance and user-response assessment: Results of an extended driver study. *Int. J. Hydrogen Energy* **2018**, *43*, 12442–12454. [CrossRef]

42. Kendall, K.; Kendall, M.; Liang, B.; Liu, Z. ScienceDirect Hydrogen vehicles in China: Replacing the Western. *Int. J. Hydrogen Energy* **2017**, *42*, 30179–30185. [CrossRef]

43. European Commission. Climate Action. Reducing CO_2 Emissions from Passenger Cars. 2017. Available online: https://ec.europa.eu/clima/policies/transport/vehicles/cars_en (accessed on 21 April 2017).

chemengineering

MDPI

Article

Hydrogen and Oxygen Evolution in a Membrane Photoreactor Using Suspended Nanosized Au/TiO$_2$ and Au/CeO$_2$

Tiziana Marino [1,*], Alberto Figoli [1], Antonio Molino [2], Pietro Argurio [3] and Raffaele Molinari [3,*]

1   Institute on Membrane Technology (ITM), National Research Council of Italy (CNR), Via P. Bucci Cubo 17C, I-87036 Rende (CS), Italy; a.figoli@itm.cnr.it
2   Research Centre of Portici, Italian National Agency for New Technologies, Energy and Sustainable Economic Development (ENEA), Piazzale Enrico Fermi 1, Portici, 80055 Napoli, Italy; antonio.molino@enea.it
3   Department of Environmental and Chemical Engineering, University of Calabria, Via P. Bucci, 44/A, I-87036 Rende (CS), Italy; pietro.argurio@unical.it
*   Correspondence: t.marino@itm.cnr.it (T.M.); raffaele.molinari@unical.it (R.M.)

Received: 8 October 2018; Accepted: 4 January 2019; Published: 10 January 2019

Abstract: Photocatalysis combined with membrane technology could offer an enormous potential for power generation in a renewable and sustainable way. Herein, we describe the one-step hydrogen and oxygen evolution through a photocatalytic membrane reactor. Experimental tests were carried out by means of a two-compartment cell in which a modified Nafion membrane separated the oxygen and hydrogen evolution semi-cells, while iron ions permeating through the membrane acted as a redox mediator. Nanosized Au/TiO$_2$ and Au/CeO$_2$ were employed as suspended photocatalysts for hydrogen and oxygen generation, respectively. The influence of initial Fe^{3+} ion concentration, ranging from 5 to 20 mM, was investigated, and the best results in terms of hydrogen and oxygen evolution were registered by working with 5 mM Fe^{3+}. The positive effect of gold on the overall water splitting was confirmed by comparing the photocatalytic results obtained with the modified/unmodified titania and ceria. Au-loading played a key role for controlling the photocatalytic activity, and the optimal percentage for hydrogen and oxygen generation was 0.25 wt%. Under irradiation with visible light, hydrogen and oxygen were produced in stoichiometric amounts. The crucial role of the couple Fe^{3+}/Fe^{2+} and of the membrane on the performance of the overall photocatalytic system was found.

Keywords: water splitting; Z-scheme; photocatalysis; photocatalytic membrane reactor

1. Introduction

Photocatalytic water splitting to generate hydrogen from solar light is a process that can play an important role for the future development of clean and renewable energies alternative to fossil fuels [1–8]. The combination of photocatalysis, which allows converting solar energy into chemical energy, and membrane-based operations could offer the possibility to achieve one-step hydrogen generation from water splitting at ambient temperature without needing further energy inputs. Hydrogen attracted increasing interest as a valid candidate for fossil-fuel substitution, enough to give rise to create the so-called "hydrogen economy" in 1970 [9–12].

From a thermodynamic point of view, hydrogen oxidation can offer three times the energy per gram in comparison to fossil fuels as gasoline. Moreover, hydrogen oxidation leads to the formation of water, making it a zero-emission fuel. Hydrogen also finds applications in the chemical industry as a reagent in fine-chemical synthesis [12].

Since Fujishima and Honda discovered photocatalytic water splitting by means of TiO_2 electrodes in 1972 [13], noticeable works were carried out in order to investigate photocatalytic-based hydrogen generation from water via both photocatalysis and photoelectrochemistry [14]. Particular attention was paid to semiconductor oxides, due to their simple preparation through calcination and their stability toward oxygen generation [15–19]. The water-splitting pathway involves a series of radical reactions initiated by light-driven photocatalyst activation, as deeply described in the literature [3].

Among the well-known semiconductors, nanosized TiO_2 is attracting more and more interest, due to its unique properties of physico-chemical stability and inertness, low cost, biocompatibility, durability, long-term photo-stability, and potent oxidative power under ultraviolet (UV)-light irradiation [20–22]. Despite its numerous advantages, the use of TiO_2 still presents a limitation, i.e., only UV light, representing ~4% of total solar energy, can be absorbed by titania nanoparticles. Therefore, extending TiO_2 light absorption to the visible fraction represents a challenging target for photocatalysis applications. Various techniques, such as noble-metal and non-metal doping, dye sensitization, and coupling with carbon materials were adopted in order to modify the electronic band of titania [23].

In a previous work, we reported that gold nanoparticles supported on titania (Au/TiO_2) represent a suitable photocatalyst for the generation of hydrogen with visible light using methanol and ethylenediaminetetraacetic acid (EDTA) as sacrificial electron donors [14]. A remarkable overall efficiency of approximately 5% measured at 560 nm was determined [14]. In this system, gold nanoparticles act as a durable and stable photosensitizer, absorbing visible light and injecting electrons in the conduction band of the TiO_2. Using 0.25 wt% gold loading on titania and methanol as a sacrificial agent, ~98 µmol of hydrogen was obtained (catalyst content 2 g/L) after 4 h of irradiation with visible light [14].

Also, we showed that ceria of small average particle size can behave as a semiconductor with a remarkably high efficiency for the photocatalytic generation of oxygen from water using Ag^+ or Ce^{3+} as a sacrificial electron acceptor [24]. The best performing CeO_2 sample was that prepared using a biopolymer "alginate" as a templating agent to synthesize about 5 nm of average particle size with a Brunauer-Emmett-Teller (BET) surface area of 93 $m^2 \cdot g^{-1}$ [24]. Also, in this case, visible-light photocatalytic activity was implemented by depositing gold nanoparticles on CeO_2 (Au/CeO_2), which allowed achieving ~50 µmol of oxygen when the aqueous suspension containing 0.25 wt% gold deposited on CeO_2 was irradiated for 4 h with visible light in the presence of $AgNO_3$ as a sacrificial agent [24].

Considering the above precedents and the visible-light photocatalytic activity for independent hydrogen and oxygen generation using Au/TiO_2 or Au/CeO_2, respectively, in the presence of an appropriate sacrificial agent, it occurs that these two photocatalysts could also work in a system to perform the simultaneous generation of hydrogen and oxygen in the absence of sacrificial agents using a Z-scheme [25–28]. In this Z-scheme methodology, hydrogen and oxygen are generated photocatalytically in different cells that are irradiated and separated by a membrane [5,29–34]. An electrolyte is used to ensure the electroneutrality in each cell and to allow charge transfer from one compartment to the other.

Li et al. [5] reported concomitant hydrogen generation and phenol degradation in a photocatalytic twin reactor under simulated solar light. In the proposed system, Pt/STO:Rh was used as a photocatalyst for hydrogen evolution, while WO_3 was chosen for phenol oxidation. A Nafion membrane separated the two compartments of the cell. Fe^{3+}/Fe^{2+} pairs were used for electron transfer. The obtained data evidenced a hydrogen generation rate of 1.90 $\mu mol \cdot g^{-1} \cdot h^{-1}$. Moreover, by employing a phenol initial concentration of 200 $\mu mol \cdot L^{-1}$, hydrogen yield reached a value of 11.37 $\mu mol \cdot g^{-1}$ after 6 h of irradiation, corresponding to an increase of 20% compared to that of pure water splitting. Fujihara et al. [35] studied water splitting in a two-compartment cell using Pt/TiO_2-anatase as a catalyst for hydrogen generation suspended in a Br_2/Br^- aqueous solution and Pt/TiO_2-rutile for simultaneous oxygen evolution in an Fe^{3+}/Fe^{2+} redox couple solution. Yu et al. [33]

also reported the experimentation of a twin reactor for the synchronized formation of hydrogen, through the photocatalytic activity of $BiVO_4$, and oxygen using $Pt/SrTiO_3$:Rh. A Nafion membrane was inserted in the two-side system to assure the separate generation of the two gases. Similarly to the above mentioned works, the Fe^{3+}/Fe^{2+} redox couple was also selected in this case as a redox mediator.

Hydrogen and oxygen in a stoichiometric ratio (2:1) were obtained, with a maximum hydrogen formation rate of 0.65 $\mu mol \cdot g^{-1} \cdot h^{-1}$. Lo et al. [31] investigated the water-splitting process for the one-step H_2 and O_2 generation by means of a membrane twin reactor under visible-light irradiation. $Pt/SrTiO_3$:Rh and WO_3 were designated as hydrogen- and oxygen-evolution semiconductors, respectively. The formation of the two gases reflected their stoichiometric ratio, with an average hydrogen evolution rate equal to 1.59 $\mu mol \cdot g^{-1} \cdot h^{-1}$.

Nafion represents a valid choice as a membrane material, since it is characterized by outstanding chemical and physical resistance [36] and an affinity for iron species. Ramirez et al. [37] investigated the uptake characteristics of different cations (Fe^{3+}, Cu^{2+}, and Ni^{2+}) by Nafion 117, which is commonly used as a separator for different chemical processes. The membrane exhibits its affinity in the order $Fe^{3+} \geq Ni^{2+} \geq Cu^{2+}$, similar to that reported in a previous study [38]. In another study [39], it was reported that a Nafion/Fe membrane was resistant to the attack of the highly oxidative radical •OH ($E°•OH/-OH = 1.90$ eV vs. normal hydrogen electrode (NHE)) and did not allow leaching of the Fe exchanged on the sulfonic groups within the 3000-h testing period. Kiwi et al. [40] demonstrated that iron compounds supported on a Nafion membrane gave good results in a photo-Fenton water treatment process, where the Fe ions were fixed and remained active in H_2O_2 decomposition.

In the present work, we studied the overall photocatalytic water splitting using Au nanoparticles as a sensitizer of TiO_2 and CeO_2 semiconductors irradiated with visible light in combination, a Nafion film as a membrane separating two cells, and ferric sulfate as the electrolyte. Diffusion tests on the iron-modified Nafion membrane were performed to determine its ability to allow diffusion of iron species. The influence on system performance of initial Fe^{3+} concentration and of Au loading on the O_2 and H_2 evolution photocatalysts was determined. Finally, the photocatalytic activity of the pairs (Au/TiO_2 and Au/CeO_2) for the photocatalytic water splitting through a Z-scheme under visible light was demonstrated.

2. Materials and Methods

TiO_2 (particle size: 20 nm; rutile/anatase: 85:15, 99.9%) was a commercial P25 sample supplied by Degussa. CeO_2 was prepared starting from an aqueous solution of alginate that was flocculated with cerium nitrate followed by calcination as previously reported [24]. Gold nanoparticles were deposited on TiO_2 and CeO_2 via the deposition–precipitation method starting from $AuHCl_4$ and maintaining pH = 10 as described in Reference [14]. TEM images were recorded with a Jeol 200 Cx microscope operating at 200 kV.

2.1. Photocataytic Tests

The photocatalytic membrane reactors (PMRs) experimented in the present work for simultaneous generation of hydrogen and oxygen from water splitting mimic the Z-scheme mechanism used by plants for natural photosynthesis. The Z-scheme (Figure 1a) basically includes two photocatalysts: (i) the oxygen evolution photocatalyst (Au/CeO_2 in this study) leading to O_2 formation via water oxidation, and (ii) the hydrogen evolution photocatalyst (Au/TiO_2 in this study) which promotes H_2 formation via water reduction. The aqueous suspensions containing the two photocatalysts need to be separated by a membrane able to transports electrons via a redox couple (Fe^{3+}/Fe^{2+} in this study) acting as an electron redox mediator.

The experimental set-up (Figure 1b) consisted of a two-compartment Pyrex cell, each with a volume of 50 mL, separated by a Nafion modified membrane with an exposed membrane surface area of 3.14 cm^2. Each compartment, containing 60 mg of photocatalyst suspended in 30 mL of Milli-Q water (corresponding to 2 $g \cdot L^{-1}$ photocatalyst concentration), was irradiated with a 125-W medium-pressure mercury lamp (DLU, HDLM E27) equipped with a Pyrex glass jacket which allows maintaining the system at a temperature of 20 °C. The suspensions were purged with argon flow for at least 30 min before irradiation in order to remove dissolved air. For polychromatic visible-light irradiation, an $Fe_2(SO_4)_3$ solution (3% w/v), circulated into the Pyrex glass jacket of the lamp, was used as a cut-off filter ($\lambda > 400$ nm). Hydrogen and oxygen generation was determined by injecting 0.1 mL of each Pyrex cell headspace gas in a gas chromatograph (GC; Agilent 7890A) equipped with a thermal conductivity detector. The GC determination was carried out by operating in isothermal conditions (50 °C), with a capillary column (CP-PoraPLOT Q, molecular sieve, 530 µm inner diameter, 15 m length) and argon as a carrier gas.

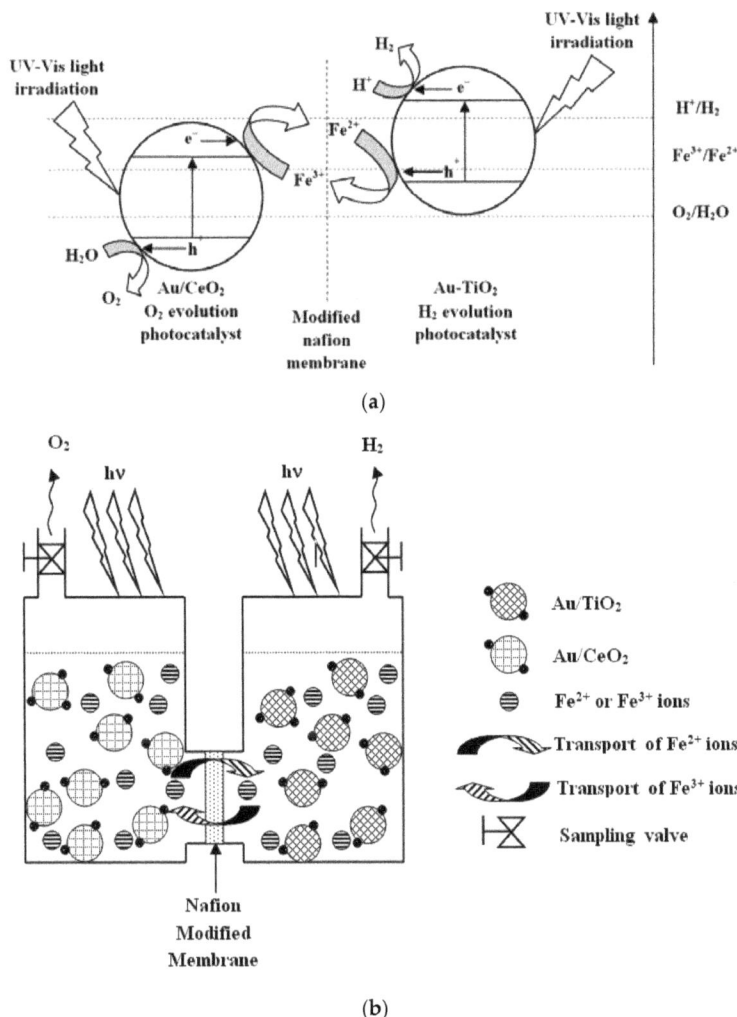

(a)

(b)

Figure 1. (**a**) Diagram of the Z-scheme overall water splitting using Au/CeO_2 as a photocatalyst for oxygen generation, Au/TiO_2 for hydrogen generation, and Fe^{3+}/Fe^{2+} as a redox couple. (**b**) Conceptual scheme of the set-up used for photocatalytic experiments.

2.2. Membrane Modification

The Nafion 117 membrane (Aldrich, thickness of 178 μm) was modified using the procedure described herein before its use. In the first step, the membrane was boiled in a 1 M HNO_3 solution for 2 h to remove any contaminant; then, it was washed with Milli-Q water, immersed in a 1 M NaOH solution, and subsequently in a 1 M H_2SO_4 solution for 4 h, each for conditioning the membrane [41]. Finally, the Nafion membrane was immersed in a 0.5 M $Fe_2(SO_4)_3$ solution for 24 h to change its functional groups (H^+) with Fe^{3+} ions.

2.3. Iron Spectrophotometric Determination

Spectrophotometric analyses were performed using 1,10-phenanthroline (Fluka, Aquanal Plus kit) and potassium thiocyanate (Aldrich), for total iron (λ = 510 nm) and ferric ion (λ = 477 nm) determinations, respectively. Ferrous iron ion concentration present in solution was calculated by the difference between total and ferric iron concentration.

3. Results and Discussion

3.1. Diffusion Test

The iron-containing modified Nafion membrane was tested to determine its ability to allow diffusion of iron species. Similar concentrations of Fe^{2+} and Fe^{3+} as that utilized in the photocatalytic experiments were used in one compartment and, then, the presence of corresponding iron species over time was determined in the other compartment. The corresponding profiles of iron species diffusing through the membrane are provided in Figure 2.

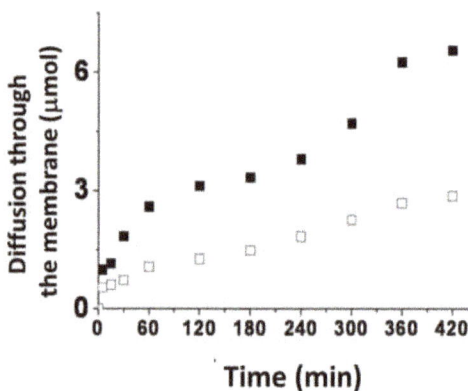

Figure 2. Fe^{2+} (■) and Fe^{3+} (□) transport over time through the Nafion membrane in the compartment cell with no initial salt.

It was observed that, while Fe^{3+} diffusion could be fitted by a straight line, i.e., the concentration of Fe^{3+} in the chamber without it grew linearly over time, similar experiments with Fe^{2+} salt clearly revealed two regimes. These results might be explained considering that, in the first regime, ion exchange of Fe^{3+} and Fe^{2+} occurred in the Nafion membrane concomitantly to the diffusion. In the second regime, the Nafion membrane behaved essentially as a Fe^{2+} exchanger.

The iron content in the membrane before and after its use was estimated by SEM energy dispersive X-ray spectroscopy (EDX) analysis. The results showed an iron weight loading of ~1.3% and this value did not change after using the membrane (Figure 3).

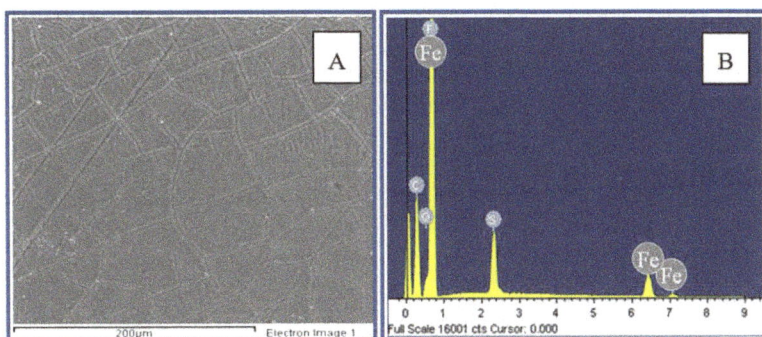

Figure 3. SEM picture of Nafion membrane used for the photocatalytic test (**A**) and energy-dispersive X-ray spectroscopy (EDX) analysis of ferric ion-modified membrane (**B**).

Figure 4 reports TEM pictures of the Au/TiO_2 and Au/CeO_2 photocatalysts. The average particle size of photocatalyst nanoparticles, determined by considering a statistically relevant number of particles in the TEM images of the samples, resulted in 2.7 and 5.0 nm for Au/TiO_2 and Au/CeO_2, respectively.

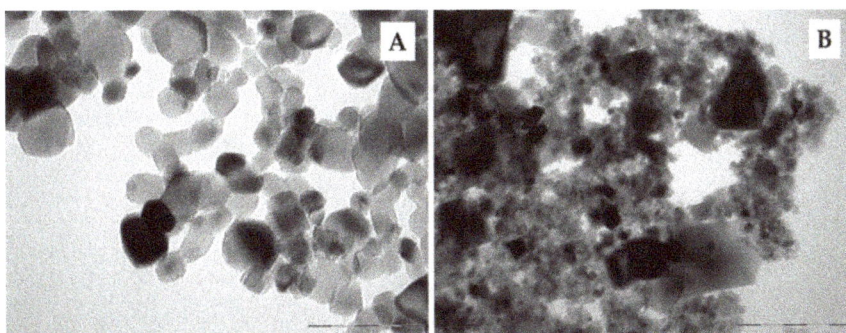

Figure 4. TEM images of Au/TiO_2 (**A**) and Au/CeO_2 (**B**).

3.2. Photocatalytic Tests

For the irradiation of the suspensions, a 125-W medium-pressure mercury lamp was used. This lamp exhibited emission peaks at a discrete wavelength, mostly in the visible region (total irradiation 2133 mW/m^2 with 61% in the visible range). The inset of Figure 5 shows the emission peaks in the spectrum of the lamp used.

Preliminary photocatalytic tests were carried out evidencing the crucial role of the couple Fe^{3+}/Fe^{2+} on both hydrogen and oxygen generation. Indeed, by starting the working of the cell with Fe^{2+} ions in the cell containing Au/TiO_2 and Fe^{3+} ions in the cell containing Au/CeO_2, oxygen generation in both cells was observed during the first 30 min. We suggest that the reason why oxygen was generated in Au/TiO_2 cell instead of the expected hydrogen was because Fe^{3+} was formed by Fe^{2+} oxidation, in the presence of oxygen by air, during the preparation of the work-up cell. Analysis of the Fe^{3+} concentration in the Au/TiO_2 cell confirmed that about 22% of Fe^{2+} ions were oxidized to Fe^{3+} during this stage. The formed Fe^{3+} quenched the generation of hydrogen and promoted the formation of oxygen, owing to the preferential reduction of Fe^{3+} by photogenerated electrons with respect to the photoreduction of water. As a consequence, during the first 30 min, oxygen generation (instead of hydrogen) was observed into the Au/TiO_2 cell before changing to continuous hydrogen generation.

On the basis of these preliminary results, to avoid the presence of Fe^{3+} ions in the Au/TiO_2 cell, the iron ions were initially added to the system only in the form of Fe^{3+} at different concentrations, as reported in Table 1, in the cell containing Au/CeO_2, while, in the Au/TiO_2 cell, the initial iron concentration was zero. Figure 5 shows the temporal evolution of hydrogen in the Au (0.25 wt%)/TiO_2 cell and oxygen in the Au (0.25 wt%)/CeO_2 cell when 5 mM Fe^{3+} was added in the Au/CeO_2 cell. Operating under these conditions, the reaction started on Au/CeO_2 with oxygen evolution via water oxidation coupled with the reduction of Fe^{3+} to Fe^{2+}. The so-produced Fe^{2+} ions permeated across the Nafion modified membrane starting hydrogen evolution in the Au/TiO_2 compartment. As a consequence, during this early stage of the reaction (approximately 5 min), the ratio $H_2:O_2$ was about 1:1 (see Figure 5) and evidenced that H_2 generation was promoted by Fe^{2+} which permeated through the membrane. After this initial stage, two regimes can be observed: a first one, until 60 min, and a second one, from 60 to 300 min. This trend can be explained considering that, in Figure 2, a similar trend with the same two regimes can be observed for Fe^{2+} permeation through the membrane. It could be deduced that Fe^{2+} ion diffusion during the first regime was faster than that during the second regime (0.043 vs. 0.022 $\mu mol_{Fe^{2+}} \cdot min^{-1}$) and this behavior affected the photocatalytic reaction. Indeed, the Fe^{2+} ions produced in the Au/CeO_2 cell permeated across the membrane and promoted a generation of hydrogen in the Au/TiO_2 cell. After 300 min, a very low hydrogen production was detected, probably because of the negligible permeation of Fe^{2+} ions across the membrane (see Figure 2) which blocked the Z-scheme mechanism. However, it should be observed that both photocatalysts efficiently worked during these two regimes (slopes of 1.25 $\mu mol_{H2} \cdot min^{-1}$ during the first regime and 0.41 $\mu mol_{H2} \cdot min^{-1}$ during the second regime), promoting the simultaneous generation of hydrogen and oxygen in a stoichiometric amount in the Au/TiO_2 and Au/CeO_2 cells, respectively, and that blocking was only caused by Fe^{2+} ion permeation through the membrane.

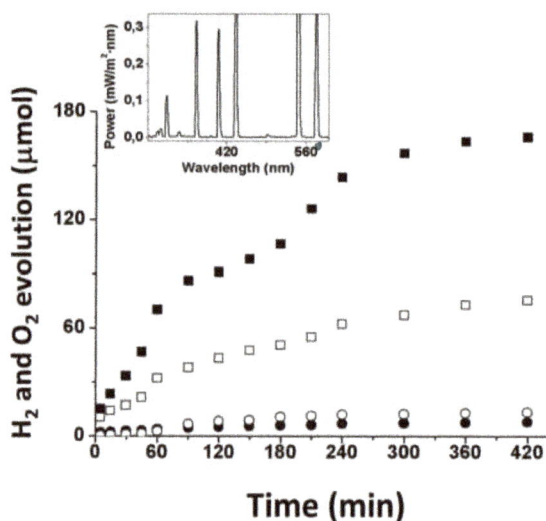

Figure 5. Hydrogen production (■) using Au (0.25 wt%)/TiO_2 and oxygen production (□) using Au (0.25 wt%)/CeO_2 compared with hydrogen production (●) using TiO_2 and oxygen production (○) using CeO_2 only (initial Fe^{3+} concentration was 5 mM in the Au/CeO_2 cell and initial Fe^{2+} concentration was zero in the Au/TiO_2 cell).

The initial concentration of Fe^{3+} ions in the Au/CeO_2 cell varied in the range from 2 to 50 mM, and it was found that 5 mM gave the best photocatalytic performance with the highest initial reaction rate (Table 1).

Table 1. Photocatalytic activity (initial reaction rate, r_0, evolved hydrogen and oxygen at 7 h) of the series of ferric aqueous solutions (initial Fe^{3+} concentration, C_0) under study, using Au (0.25 wt%)/TiO$_2$ and Au (0.25 wt%)/CeO$_2$ for hydrogen and oxygen generation, respectively.

C_0 Fe^{3+} (mM)	Evolved H$_{2,7h}$ (μmol)	Evolved O$_{2,7h}$ (μmol)	H$_2$ $r_0 \times 10^2$ (μmol$\cdot$min^{-1})	O$_2$ $r_0 \times 10^2$ (μmol$\cdot$min^{-1})
2	56.2	28.1	34.2	20.7
5	166.1	75.6	114.0	53.4
10	86.2	43.0	96.3	32.5
20	43.2	24.2	34.8	21.1
50	25.5	16.3	33.4	23.5

Au/CeO$_2$ was selective for oxygen generation as it was found to be unable to generate hydrogen. The reason for this behavior might be related to the energy of the electrons in the conduction band of Au/CeO$_2$, which was not sufficient for water reduction (see Figure 1). In contrast, these electrons were able to reduce Fe^{3+} to Fe^{2+}.

The influence of gold on the photocatalytic performance of the Z-scheme overall water splitting by TiO$_2$ and CeO$_2$ was determined by comparing the photocatalytic activity of the gold-containing semiconductors in comparison with the same semiconductors without gold (Figure 5).

As expected, despite the irradiation source used not permitting exclusively visible-light emission, a positive effect of the presence of gold was observed.

The gold loading plays an important role in the photocatalytic efficiency of the system. To demonstrate this point, we performed some photocatalytic tests with equal gold loading on titania and ceria varying from 0.25 to 1.0 wt%. By measuring the initial rate of hydrogen generation, it was concluded that the optimum gold loading under these conditions was the lowest (0.25 wt%) (Table 2).

Table 2. Photocatalytic activity (initial reaction rate, r_0, evolved hydrogen and oxygen at 7 h) of the series of gold-containing samples under study, using a 5 mM ferric solution and TiO$_2$ and CeO$_2$ as photocatalysts for hydrogen and oxygen generation, respectively.

Au Loading (wt%)	Evolved H$_{2,7h}$ (μmol)	Evolved O$_{2,7h}$ (μmol)	H$_2$ $r_0 \times 10^2$ (μmol$\cdot$min^{-1})	O$_2$ $r_0 \times 10^2$ (μmol$\cdot$min^{-1})
0.25	166.1	75.6	114.0	53.4
0.6	152.0	61.3	112.1	70.8
1.0	71.0	25.1	67.0	32.2

Au content in the aqueous reacting environment after the reaction was below the detection limit (0.1 ppm).

In a precedent work it was also observed that Au loading is a key parameter controlling the photocatalytic activity of Au/TiO$_2$ and that an optimal Au percentage exists. Since the presence of Au nanoparticles is detrimental for the efficiency of TiO$_2$ photocatalysis under UV irradiation, but enhances the photocatalytic efficiency for visible-light irradiation, the observed beneficial effect of Au might be explained considering that, by operating under our conditions, visible light gave the largest contribution to the total photocatalytic water splitting. In view of the above data, we propose the water-splitting mechanism shown in Figure 6.

Upon light absorption (mostly visible), electrons (in the conduction band) and holes (in the valence band) were generated in Au/CeO$_2$. It was found that electrons in Au/CeO$_2$ were inefficient to generate hydrogen and, therefore, they were captured by Fe^{3+}, forming Fe^{2+} ions that diffused through the Nafion membrane to the Au/TiO$_2$ cell. The holes located on Au of the Au/CeO$_2$ photocatalyst had sufficient oxidation power to generate oxygen via water oxidation as previously proven [42]. In the Au/TiO$_2$ cell, similar charge separation as in the Au/CeO$_2$ cell would occur upon photon absorption; however, in this case, the reduction potential of the TiO$_2$ conduction band had enough energy to form hydrogen via water reduction. The holes on Au were, in this case, quenched by oxidation of Fe^{2+} diffusing through the Nafion membrane from the Au/CeO_2 cell.

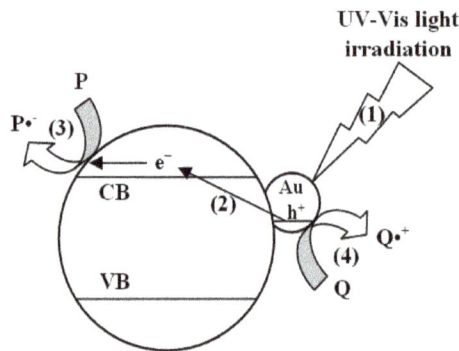

Figure 6. Elementary steps occurring in the photocatalytic reaction upon irradiation of a gold-containing photocatalyst sample under ultraviolet (UV) irradiation: (1) photon absorption; (2) electron migration from Au to semiconductor conduction band; (3) electron quenching by P; (4) Q oxidation by H^+ mediated by Au.

To demonstrate the photocatalytic activity of the pair Au/TiO_2 and Au/CeO_2 for the overall water splitting through a Z-scheme under visible light, analogous photocatalytic experiments were performed by filtering light to almost completely remove the irradiation wavelength below 400 nm (Figure 7). Under these conditions, the irradiance decreased from 2133 $mW \cdot m^{-2}$ to 570 $mW \cdot m^{-2}$. In Figure 7, the two regimes previously observed in Figure 5 were not present. This can be ascribed to the lower irradiance (3.6 times lower for the visible light compared to UV irradiation), which decreased the H_2/O_2 generation rate (slopes of 0.08 $\mu mol_{H2} \cdot min^{-1}$ and 0.04 $\mu mol_{O2} \cdot min^{-1}$), thus requiring a lower Fe^{2+} ion diffusion rate.

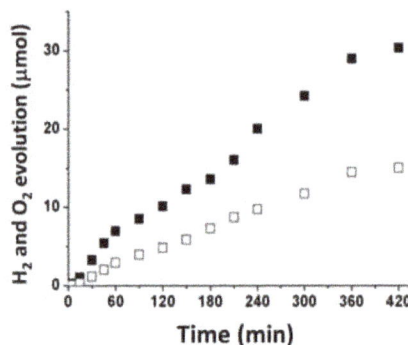

Figure 7. Hydrogen production (■) using Au (0.25 wt%)/TiO_2 and oxygen production (□) using Au (0.25 wt%)/CeO_2, under visible-light irradiation (initial Fe^{3+} concentration was 5 mM in the Au/CeO_2 cell and initial Fe^{2+} concentration was zero in the Au/TiO_2 cell).

Also under these conditions, hydrogen and oxygen generation was observed in stoichiometric amounts in the Au/TiO_2 and Au/CeO_2 cells, respectively (Table 3).

Table 3. Photocatalytic activity (initial reaction rate, r_0, evolved hydrogen and oxygen at 7 h) of the Au (0.25 wt%)/TiO$_2$ and Au (0.25 wt%)/CeO$_2$ samples using a 5 mM ferric solution for hydrogen and oxygen generation, respectively, under visible-light irradiation.

Au Loading (wt%)	Evolved H$_{2,7h}$ (µmol)	Evolved O$_{2,7h}$ (µmol)	H$_2$ $r_0 \times 10^2$ (µmol·min^{-1})	O$_2$ $r_0 \times 10^2$ (µmol·min^{-1})
0.25	30.36	14.89	11.0	4.2

4. Conclusions

In the present article, we combined the high activity of Au/TiO$_2$ to photocatalytically generate hydrogen with the high activity of Au/CeO$_2$ to promote oxygen evolution from water. Thus, overall water splitting was obtained by operating through a Z-scheme using a Nafion membrane and Fe^{3+}/Fe^{2+} as an electrolyte.

The obtained results showed the following:

— The employed photocatalysts promoted simultaneous hydrogen and oxygen generation;
— The optimal content of ferric ions in the Au/CeO$_2$ compartment was 5 mM;
— Gold operated as a photosensitizer allowing photocatalytic hydrogen and oxygen formation under visible light;
— A gold loading of 0.25 wt% led to the best results in terms of hydrogen and oxygen evolution (166.1 and 75.6 µmol, respectively, after 7 h of UV-visible-light irradiation);
— Hydrogen and oxygen were produced in stoichiometric amounts, i.e., 30.36 and 14.89 µmol, respectively, after 7 h of irradiation with visible light;
— The decrease in permeation rate of iron ions through the Nafion membrane affected the photocatalytic performance, slowing the generation rates of both hydrogen and oxygen.

Author Contributions: T.M. and R.M. conceived and designed the experiments; T.M. performed the experiments; T.M., R.M., P.A. and A.M. analyzed the data; R.M. and P.A. contributed reagents/materials/analysis tools; T.M., R.M., P.A., A.M., A.F. wrote the paper.

Funding: This research received no external funding.

Conflicts of Interest: The authors declare no conflicts of interest.

References

1. Bamwenda, G.R.; Tsubota, S.; Nakamura, T.; Haruta, M. Photoassisted Hydrogen-Production from a water-ethanol solution—A Comparison of activities of Au-TiO$_2$ and Pt-TiO$_2$. *J. Photochem. Photobiol. A-Chem.* **1995**, *89*, 177–189. [CrossRef]
2. Khan, S.U.M.; Al-Shahry, M.; Ingler, W.B. Efficient photochemical water splitting by a chemically modified n-TiO$_2$. *Science* **2002**, *297*, 2243–2245. [CrossRef] [PubMed]
3. Loddo, V.; Palmisano, L.; Marino, T.; Molinari, R. Membranes for photocatalysis in water and wastewater treatment. In *Advanced Membrane Science and Technology for Sustainable Energy and Environmental Applications*; Basile, A., Nunes, S.P., Eds.; Woodhead Publishing: Sawston, UK, 2011; pp. 746–768.
4. Maeda, K.; Domen, K. New non-oxide photocatalysts designed for overall water splitting under visible light. *J. Phys. Chem. C* **2007**, *111*, 7851–7861. [CrossRef]
5. Li, D.; Yu, J.C.; Nguyen, V.; Wu, C.S.; Wang, X. A dual-function photocatalytic system for simultaneous separating hydrogen from water splitting and photocatalytic degradation of phenol in a twin-reactor. *Appl. Catal. B Environ.* **2018**, *239*, 268–279. [CrossRef]
6. Argurio, P.; Fontananova, E.; Molinari, R.; Drioli, E. Photocatalytic Membranes in Photocatalytic Membrane Reactors. *Processes* **2018**, *6*, 162. [CrossRef]
7. Iglesias, O.; Rivero, M.J.; Urtiaga, A.M.; Ortiz, I. Membrane-based photocatalytic systems for process intensification. *Chem. Eng. J.* **2016**, *305*, 136–148. [CrossRef]

8. Rahimpour, M.R.; Mahmoodi, L. *Performance of Reactors With PMs*; Elsevier Inc.: Amsterdam, The Netherlands, 2018; ISBN 9780128135495.

9. Molino, A.; Migliori, M.; Blasi, A.; Davoli, M.; Marino, T.; Chianese, S.; Catizzone, E.; Giordano, G. Municipal waste leachate conversion via catalytic supercritical water gasification process. *Fuel* **2017**, *206*, 155–161. [CrossRef]

10. Marbán, G.; Valdés-Solís, T. Towards the hydrogen economy? *Int. J. Hydrog. Energy* **2007**, *32*, 1625–1637. [CrossRef]

11. Armaroli, N.; Balzani, V. The hydrogen issue. *ChemSusChem* **2011**, *4*, 21–36. [CrossRef]

12. Ball, M.; Wietschel, M. The future of hydrogen—Opportunities and challenges. *Int. J. Hydrog. Energy* **2009**, *34*, 615–627. [CrossRef]

13. Fujishima, A.; Honda, K. Electrochemical photolysis of water at a semiconductor electrode. *Nature* **1972**, *238*, 37–38. [CrossRef] [PubMed]

14. Silva, C.G.; Juárez, R.; Marino, T.; Molinari, R.; García, H. Influence of excitation wavelength (UV or visible light) on the photocatalytic activity of titania containing gold nanoparticles for the generation of hydrogen or oxygen from water. *J. Am. Chem. Soc.* **2011**, *133*, 595–602. [CrossRef] [PubMed]

15. Galiano, F.; Song, X.; Marino, T.; Boerrigter, M.; Saoncella, O.; Simone, S.; Faccini, M.; Chaumette, C.; Drioli, E.; Figoli, A. Novel Photocatalytic PVDF/Nano-TiO$_2$ Hollow Fibers for Environmental Remediation. *Polymers* **2018**, *10*, 1134. [CrossRef]

16. Molinari, R.; Caruso, A.; Palmisano, L. Photocatalytic Processes in Membrane Reactors. In *Comprehensive Membrane Science and Engineering*; Drioli, E., Giorno, L., Eds.; Elsevier Science B.V.: Oxford, UK, 2010; Volume 3, pp. 165–193, ISBN 9780080932507.

17. Hoffmann, M.R.; Martin, S.T.; Choi, W.; Bahnemann, D.W. Environmental Applications of Semiconductor Photocatalysis. *Chem. Rev.* **1995**, *95*, 69–96. [CrossRef]

18. Schneider, J.; Matsuoka, M.; Takeuchi, M.; Zhang, J.; Horiuchi, Y.; Anpo, M.; Bahnemann, D.W. Understanding TiO$_2$ Photocatalysis: Mechanisms and Materials. *Chem. Rev.* **2014**, *114*, 9919–9986. [CrossRef] [PubMed]

19. Kudo, A.; Miseki, Y. Heterogeneous photocatalyst materials for water splitting. *Chem. Soc. Rev.* **2009**, *38*, 253–278. [CrossRef] [PubMed]

20. Marino, T.; Boerigter, M.; Faccini, M.; Chaumette, C.; Arockiasamy, L.; Bundschuh, J.; Figoli, A. Photocatalytic activity and synthesis procedures of TiO$_2$ nanoparticles for potential applications in membranes. In *Application of Nanotechnology in Membranes for Water Treatment*; Hoinkis, J., Figoli, A., Altinkaya, S.A., Bundschuh, J., Eds.; CRC, Taylor and Francis Group: Boca Raton, FL, USA, 2017; ISBN 9781138896581.

21. Figoli, A.; Marino, T.; Simone, S.; Boerighter, M.; Faccin, M.; Chaumette, C.; Drioli, E. Application of nano-sized TiO$_2$ in membrane technology. In *Application of Nanotechnology in Membranes for Water Treatment*; Hoinkis, J., Figoli, A., Altinkaya, S.A., Bundschuh, J., Eds.; CRC, Taylor and Francis Group: Boca Raton, FL, USA, 2017; ISBN 9781138896581.

22. Fujishima, A.; Rao, T.N.; Tryk, D.A. Titanium dioxide photocatalysis. *J. Photochem. Photobiol. C Photochem. Rev.* **2000**, *1*, 1–21. [CrossRef]

23. Kumar, S.G.; Devi, L.G. Review on modified TiO$_2$ photocatalysis under UV/visible light: Selected results and related mechanisms on interfacial charge carrier transfer dynamics. *J. Phys. Chem. A* **2011**, *115*, 13211–13241. [CrossRef]

24. Primo, A.; Marino, T.; Corma, A.; Molinari, R.; Garcia, H. Efficient visible-light photocatalytic water splitting by minute amounts of gold supported on nanoparticulate CeO$_2$ obtained by a biopolymer templating method. *J. Am. Chem. Soc.* **2011**, *133*, 6930–6933. [CrossRef]

25. Marino, T.; Primo, A.; Corma, A.; Molinari, R.; Garcia, H. Photocatalytic overall water splitting by combining gold nanoparticles supported on TiO$_2$ and CeO$_2$. In Proceedings of the 2nd European Symposium on Photocatalysis, Bordeaux, France, 19 October 2011; p. 76.

26. Higashi, M.; Abe, R.; Ishikawa, A.; Takata, T.; Ohtani, B.; Domen, K. Z-scheme Overall Water Splitting on Modified-TaON Photocatalysts under Visible Light (λ < 500 nm). *Chem. Lett.* **2008**, *37*, 138–139. [CrossRef]

27. Maeda, K.; Higashi, M.; Lu, D.; Abe, R.; Domen, K. Efficient nonsacrificial water splitting through two-step photoexcitation by visible light using a modified oxynitride as a hydrogen evolution photocatalyst. *J. Am. Chem. Soc.* **2010**, *132*, 5858–5868. [CrossRef] [PubMed]

28. Chiarello, G.L.; Tealdi, C.; Mustarelli, P.; Selli, E. Fabrication of Pt/Ti/TiO$_2$ Photoelectrodes by RF-Magnetron Sputtering for Separate Hydrogen and Oxygen Production. *Materials* **2016**, *9*, 279. [CrossRef] [PubMed]

29. Selli, E.; Chiarello, G.L.; Quartarone, E.; Mustarelli, P.; Rossetti, I.; Forni, L. Photocatalytic water splitting device for separate hydrogen and oxygen evolution. *Chem. Commun.* **2007**, *47*, 5022–5024. [CrossRef] [PubMed]

30. Xu, Q.; Zhang, L.; Yu, J.; Wageh, S.; Al-ghamdi, A.A.; Jaroniec, M. Direct Z-scheme photocatalysts: Principles, synthesis, and applications. *Mater. Today* **2018**, *21*, 1042–1063. [CrossRef]

31. Lo, C.; Huang, C.; Liao, C.; Wu, J.C.S. Novel twin reactor for separate evolution of hydrogen and oxygen in photocatalytic water splitting. *Int. J. Hydrog. Energy* **2010**, *35*, 1523–1529. [CrossRef]

32. Seger, B.; Kamat, P.V. Fuel cell geared in reverse: Photocatalytic hydrogen production using a TiO$_2$/Nafion/Pt membrane assembly with no applied bias. *J. Phys. Chem. C* **2009**, *113*, 18946–18952. [CrossRef]

33. Yu, S.-C.; Huang, C.-W.; Liao, C.-H.; Wu, J.C.S.; Chang, S.-T.; Chen, K.-H. A novel membrane reactor for separating hydrogen and oxygen in photocatalytic water splitting. *J. Membr. Sci.* **2011**, *382*, 291–299. [CrossRef]

34. Kitano, M.; Tsujimaru, K.; Anpo, M. Decomposition of water in the separate evolution of hydrogen and oxygen using visible light-responsive TiO$_2$ thin film photocatalysts: Effect of the work function of the substrates on the yield of the reaction. *Appl. Catal. A Gen.* **2006**, *314*, 179–183. [CrossRef]

35. Fujihara, K.; Ohno, T.; Matsumura, M. Splitting of water by electrochemical combination of two photocatalytic reactions on TiO$_2$ particles. *J. Chem. Soc. Faraday Trans.* **1998**, *94*, 3705–3709. [CrossRef]

36. Kraytsberg, A.; Yair, E.-E. Review of Advanced Materials for Proton Exchange Membrane Fuel Cells. *Energy Fuels* **2014**, *28*, 7303–7330. [CrossRef]

37. Ramírez, J.; Godínez, L.A.; Méndez, M.; Meas, Y.; Rodriguez, F.J. Heterogeneous photo-electro-Fenton process using different iron supporting materials. *J. Appl. Electrochem.* **2010**, *40*, 1729–1736. [CrossRef]

38. Feng, J.; Hu, X.; Yue, P.L.; Zhu, H.Y.; Lu, G.Q. Discoloration and mineralization of Reactive Red HE-3B by heterogeneous photo-Fenton reaction. *Water Res.* **2003**, *37*, 3776–3784. [CrossRef]

39. Sabhi, S.; Kiwi, J. Degradation of 2,4-dichlorophenol by immobilized iron catalysts. *Water Res.* **2001**, *35*, 1994–2002. [CrossRef]

40. Kiwi, J. Innovative immobilized Fenton systems useful in the abatement of industrial pollutants. In Proceedings of the Third Asia-Pacific Conference on Sustainable Energy and Environmental Technologies, Hong Kong, China, 3–6 December 2000; p. 562.

41. Goswami, A.K.; Acharya, A.; Pandey, A.K. Study of self-diffusion of monovalent and divalent cations in Nafion-117 ion-exchange membrane. *J. Phys. Chem. B* **2001**. [CrossRef]

42. Silva, C.G.; Bouizi, Y.; Fornés, V.; Garcia, H. Layered Double Hydroxides as Highly Efficient Photocatalysts for Visible Light Oxygen Generation from Water. *J. Am. Chem. Soc.* **2009**, *131*, 13833–13839. [CrossRef] [PubMed]

chemengineering

MDPI

Article

Experimental Investigation of the Gas/Liquid Phase Separation Using a Membrane-Based Micro Contactor

Kay Marcel Dyrda *, Vincent Wilke, Katja Haas-Santo and **Roland Dittmeyer**

Karlsruhe Institute of Technology, Institute for Micro Process Engineering,
76344 Eggenstein-Leopoldshafen, Germany; vincentwilke101@gmail.com (V.W.);
katja.haas-santo@kit.edu (K.H.-S.); roland.dittmeyer@kit.edu (R.D.)
* Correspondence: kay.dyrda@kit.edu; Tel.: +49-721-608-23430

Received: 28 September 2018; Accepted: 6 November 2018; Published: 14 November 2018

Abstract: The gas/liquid phase separation of CO_2 from a water-methanol solution at the anode side of a μDirect-Methanol-Fuel-Cell (μDMFC) plays a key role in the overall performance of fuel cells. This point is of particular importance if the μDMFC is based on a "Lab-on-a-Chip" design with transient working behaviour, as well as with a recycling and a recovery system for unused fuel. By integrating a membrane-based micro contactor downstream into the μDMFC, the efficient removal of CO_2 from a water-methanol solution is possible. In this work, a systematic study of the separation process regarding gas permeability with and without two-phase flow is presented. By considering the μDMFC working behaviour, an improvement of the overall separation performance is pursued. In general, the gas/liquid phase separation is achieved by (1) using a combination of the pressure gradient as a driving force, and (2) capillary forces in the pores of the membrane acting as a transport barrier depending on the nature of it (hydrophilic/hydrophobic). Additionally, the separation efficiency, pressure gradient, orientation, liquid loss, and active membrane area for different feed inlet temperatures and methanol concentrations are investigated to obtain an insight into the separation process at transient working conditions of the μDMFC.

Keywords: gas/liquid separation; micro direct methanol fuel cell (μDMFC); porous membranes; micro channel; two-phase flow; micro contactor; separator

1. Introduction

In recent years, the operating time and the miniaturisation of portable electronic devices such as smartphones, laptops, and many others devices has become a field of high importance. Unfortunately, due to the low energy capacity of conventional electrochemical accumulators e.g., lithium ion batteries, this kind of energy supply or recharging systems (power banks) are no longer appropriate for long operating times. To overcome short operating times, many possibilities are currently being investigated to replace lithium ion battery-based energy supplies or recharging systems [1–3]. Small-sized membrane-based fuel cell systems with a highly specific energy density are among the most promising candidates to overcome short operating times. Using a micro direct methanol fuel cell (μDMFC) based on a lab-on-a-chip (LOC) design, integrated fuel supply for methanol, and a separation system, the operating time can be extended significantly without increasing the weight or volume of the portable device disproportionately. In addition, the μDMFC can be refuelled instantly with methanol, which is an easily storable, convenient liquid fuel with an energy density of 4.42 kWh/dm^3 [1–5].

During the operation of the μDMFC, methanol from the liquid water-methanol mixture is oxidised to gaseous carbon dioxide while at the cathode side, oxygen from gaseous air is reduced to liquid water. Subsequently, two-phase flows are formed by the produced carbon dioxide and water at both -anode and cathode- side. This occurring two-phase flow, on the one hand, decreases the accessibility of methanol and oxygen to the catalyst layer on the respective electrode area inside the µDMFC, leading, on the other hand, to a channel blocking phenomena in one or more channels during operation. In both cases, the two-phase flow causes a reduction of the overall fuel cell performance. Therefore, to ensure the permanent operation of a µDMFC in recirculation mode and to improve the fuel cell performance stability, produced carbon dioxide as waste product or byproduct must be removed continuously from the anodic loop. Furthermore, the water must be efficiently recovered from the cathodic loop and fed into the anodic loop to compensate the water losses during µDMFC operation and CO_2 separation [6–8].

In conventionally-operated DMFCs, the occurring two-phase flow is passively separated in small vessels due to the difference in density and the use of gravity [4]. For µDMFCs based on LOC design, the separation of a two-phase flow is significantly more complex. Only flat separation systems to realise a compact and portable design can be considered. By combining membrane- and micro-engineering -technology, an orientation-independent removal and recovery of carbon dioxide and water from the two-phase flow is possible [9–12]. This can be achieved by using a porous membrane-based micro contactor installed downstream of the µDMFC with additional integration on the bipolar plates to achieve a compact and flat LOC design including a recycling system. Different design approaches and investigations for several microcontactors are summarised in Table 1.

Table 1. Examples of previous works on the gas/liquid separation in microstructured devices using polymeric membranes or inorganic microsieves as a separation layer.

Investigator	Research Topic	Membrane	Material	Fluids	Mode
Meng et al. [9]	Distributed Breather	Microsieve	Silicon	CO_2 (g)/H_2O (l)	P
Lee et al. [10]	Micro Bubble Separator	Microsieve	Silicon	CO_2 (g)/H_2O (l)	P
Amon et al. [11]	Micro-electro-mechanical based µDMFC	Microsieve	Silicon	CO_2/H_2O + CH_3OH (l)	P
Alexander et al. [12]	Micro-breather (heat sink)	Microsieve	Silicon	H_2O (g)/H_2O (l)	P
Kraus et al. [13]	Orientation indipendent microseparator	Membrane	PTFE	CO_2 (g)/H_2O (l)	A
Meng et al. [14–16]	Membrane based micro separator in a µDMFC	Membrane	PTFE∣PP	CO_2 (g)/H_2O + CH_3OH (l)	A
Xu et al. [17]	Active gas/liquid phase separation	Membrane	ACP	N_2 (g)/H_2O (l)	A
David et al. [18]	Micro heat exchanger and microgas separator	Membrane	PTFE	Air (g)/H_2O (l)	A
Fazeli et al. [19]	Differential pressure on the gas/liquid separation	Membrane	PTFE	CO_2 (g)/H_2O + CH_3OH (l)	A

PTFE: Polytetrafluoroethylene; PP: Polypropylene; ACP: Acrylic Copolymer; P: Passive; A: Active.

All of the studies summarised in Table 1 describe different membrane-based micro contactors for the gas/liquid phase separation, studying different types of membranes and parameters. Nevertheless, these studies fail to systematically investigate the gas permeability and influence of diffusion fluxes at single and two-phase flow conditions, the separation efficiency regarding the gas permeability, and the active membrane area during the gas/liquid separation process. In this work, a membrane-based micro contactor for the gas/liquid phase separation with its corresponding systematic study of the gas permeability, influence of the diffusion fluxes, and separation performance at single and two-phase flow conditions, is presented for four different membranes.

2. Theoretical Background

To provide a solid background on the separation process in a porous membrane-based micro contactor, capillary forces and mass transfer concepts—i.e., fundamentals of passive or active gas/liquid phase separation—will be introduced.

2.1. Capillary Pressure

The capillary pressure (capillary forces), which has been known since 1830, can be mathematically described by the Young-Laplace equation for ideal circular pores Equation (1) and rectangular/square micro channels Equation (2). Detailed information about the capillary force can be found in the supplementary information.

Circular Pore

$$P_{cap} = \frac{4 \cdot \sigma_{lg} \cdot \cos(\varphi)}{d_h} \tag{1}$$

Rectangular/Square Micro Channel

$$P_{cap} = \sigma_{lg} \cdot \left(\frac{\cos(\varphi_t)}{h} + \frac{\cos(\varphi_b)}{h} + \frac{\cos(\varphi_l)}{w} + \frac{\cos(\varphi_r)}{w} \right) \tag{2}$$

where σ_{lg} represents the surface tension, φ the contact angle, and d_h the characteristic diameter (hydraulic or circular) of the pores of the polymer based membrane. Using Equation (1), the capillary pressure can be calculated indicating the maximum water entry pressure of a porous membrane. For an ideal circular pore diameter of 500 nm and a water contact angle of 100° ($\sigma_{H_2O} = 0.072$ N/m), a capillary pressure, or in a broader sense, a capillary force of −1 bar (N/m) acting against the pore wetting should be applied. The negative value of the calculated capillary pressure indicates the acting direction of the force.

Figure 1 shows the capillary pressure calculated by Equation (2) for four different micro channel configurations and pores (round capillary) depending on a scaling factor s dividing the starting values width and/or height of the channel geometry.

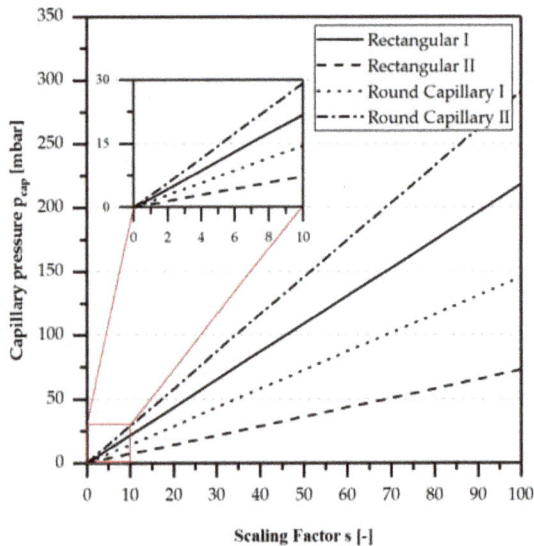

Figure 1. Capillary Pressure p_{cap} for water at 20 °C, $\sigma_{H_2O} = 0.07275$ N/m, dependent on channel geometry, surface properties, and material combination. Rectangular I (wall contact angle: 60°, width: 1 mm, height: 0.5 mm) without a hydrophobic membrane on the top of the micro channel. Rectangular II (wall contact angle: 60°, width: 1 mm, height: 0.5 mm) with hydrophobic membrane made of PTFE (contact angle: 120° [20]) on the top of the micro channel. Round Capillary I (wall contact angle: 60°, diameter: 1 mm). Round Capillary II (wall contact angle: 60°, diameter: 0.5 mm).

In Figure 1, it is clearly visible that homogenous wetted micro channels with a width of 1 mm and a height of 0.5 mm always have the highest positive capillary pressure with an increasing scaling factor. Heterogeneous wetted micro channels, as they can occur in a membrane-based micro contactor (see Figure 2), instead have the lowest positive capillary pressure due to the high influence of the hydrophobic membrane. To utilise the capillary pressure as a driving force for the passive gas/liquid phase separation, the size of the channel plays a major role and must be taken into account if passively-operated gas/liquid separation is the main objective.

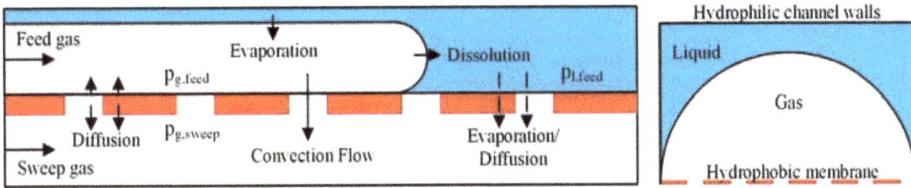

Figure 2. Cross and longitudinal section of the separation process.

However, to ensure the portability of portable energy systems or applications, the gas/liquid phase separation must be independent of orientation, which can be ensured by the capillary forces in the micro channels. In macroscopic channels, gravitational forces have a considerable influence on the behaviour of the two-phase flow (TPF), since buoyancy forces exceed the capillary forces. The influence of capillary forces increases with decreasing channel diameter, as shown in Figure 1. To quantify the ratio of capillary and gravitational forces, the Confinement number (Co) is calculated by Equation (3). The Eötvös number (Eo), related to the Confinement number, can be calculated by Equation (4).

$$Co = \frac{1}{d_h} \sqrt{\frac{\sigma_{lg}}{g \cdot (\rho_l - \rho_g)}} \tag{3}$$

$$E\ddot{o} = \frac{1}{8 \cdot Co^2} \tag{4}$$

where d_h represents the hydraulic diameter, g the gravitational acceleration, ρ_l the liquid density, and ρ_g the gas density. For a rectangular or square channel, the hydraulic diameter can be calculated using the cross-section area A and wetted perimeter U_w, as shown in Figure 1, by using the following equation:

$$d_h = \frac{4 \cdot A}{U_w} \tag{5}$$

According to Huh et al. [21], Serizawa et al. [22], Suo et al. [23] and Brauner et al. [24], the capillary forces become predominant over the buoyancy forces if $Co \geq 3.3$ or $E\ddot{o} < 0.01$. Using the data for a micro channel with a height of 0.5 mm and a width of 1.0 mm with water + air at 20 °C, a Confinement number of $Co \approx 4$ and an Etövös Number of $E\ddot{o} \approx 0.0075$ are obtained. In both cases, the capillary forces are predominant over the gravitational forces, and an orientation-independent separation can be assumed.

2.2. Mass Transfer

The working principle of the separation process is illustrated in Figure 2, showing the cross sectional (left side) and longitudinal section (right side) views of the feed and permeate channel. As can be observed, the combination of heterogeneous channel walls and hydrophobic porous membranes leads to a spatial separation behaviour of the two-phase due to the capillary forces.

In addition, it can be seen that the gas phase directly contacts the porous membrane, and is then pushed through. The driving force of the gas separation is the transmembrane pressure difference Δp_{TM}:

$$\Delta p_{TM} = p_{g,feed} - p_{g,sweep} \tag{6}$$

Furthermore, as shown in Figure 1, the capillary pressure on a gas bubble in a rectangular microchannel with a hydrophobic membrane at a reduction factor of 1 (Figure 1) is about 1 mbar. For the effective separation of typical amounts of gas over a small membrane area, higher pressures are necessary. Therefore, the capillary pressure is considered negligible and the pressure in the gas and liquid phases of the feed channel can be assumed to be equal, as shown in Equation (7).

$$P_{g,feed} = P_{l,feed} \tag{7}$$

In the longitudinal section of the micro contactor (Figure 2), an overview of the different mass transfer processes is given. Gaseous species are transported by convection as well as diffusion.

2.2.1. Convective Mass Transfer

The predominant mass transport through the gas covered area of the membrane is the pressure driven convective flow. Depending on the thermodynamic state of the system, the feed gas is saturated by the liquid species through evaporation, which is subsequently also separated from the liquid stream convectively.

In the 19th century, Henry Darcy described the laminar convectional flow through porous media empirically. His mathematical correlation is today known as the Darcy's law. As assumed by this law, the amount of the transferred fluid is directly proportional to the pressure difference across the porous structure, as described by Equation (8) [17,18]:

$$\dot{V} = \frac{\kappa \cdot A \cdot \Delta p_{TM}}{\eta \cdot l} \tag{8}$$

where $\dot{V}$ represent the volume flow rate of a fluid through a porous structure, κ the permeability of the porous structure, A the available separation area which is usually completely covered by the separable fluid, Δp_{TM} the transmembrane pressure difference across the porous structure, η the viscosity of the fluid, and l the thickness of the separation layer.

The available or active membrane area during the separation process of a two-phase flow is defined by the percentage of the overall allocable membrane area where the separable gas flow is concentrated, and only thereby available for the separation process. For determination of the active membrane area, Equation (9) can be used.

$$A_{active} = A_{mem} \cdot \frac{\kappa_{TPF}}{\kappa_g} \tag{9}$$

where A_{active} represent the active membrane area, A_{mem} the maximum available membrane area in the separation unit, κ_{TPF} the measured gas permeability with two-phase flow, and κ_g the measured gas permeability with single-phase flow.

2.2.2. Diffusive Mass Transfer

In tight porous media, the interaction between the diffusing molecules and the channel walls becomes more relevant at a Knudsen number of Kn > 0.01. If the Knudsen number exceeds this value, the diffusion regime turns to a Knudsen diffusion regime in which collisions with walls occur more often than collisions among molecules. In general, the Knudsen number is defined as:

$$Kn = \frac{\lambda}{d_p} \tag{10}$$

with

$$\lambda = \frac{4 \cdot k_B \cdot T}{\pi \cdot \sigma_{coll}^2 \cdot P} \tag{11}$$

where λ is the free mean path length of the molecule, d_P is the pore diameter, k_B is the Boltzmann constant, σ_{coll} is the collision diameter of the molecule, and T and p are the temperature and pressure respectively. At Kn > 2, only Knudsen diffusion is occurring. The Knudsen diffusion coefficient is given by [25]:

$$D_{Kn} = \frac{1}{3} \cdot d_p \cdot \sqrt{\frac{2 \cdot k_B \cdot T}{m_{mol}}} \tag{12}$$

where m_{mol} is the molecular mass. In contrast, at Kn < 0.01, free molecular diffusion can be assumed. In that case, the diffusion coefficient $D_{AB,mol}$ of substance A in substance B can be calculated using Fuller's equation with sufficient accuracy [26].

$$D_{AB,mol} = \frac{0.001 \cdot T^{1.75} \cdot \sqrt{\left(\frac{1}{M_A} + \frac{1}{M_B}\right)}}{p \cdot \left(\sqrt[3]{(v_A + v_B)}\right)^2} \tag{13}$$

where T is the temperature, M is the molar mass, p the pressure, and v is the molecular diffusion volume. To describe the diffusive mass transfer processes in the porous membrane-based micro contactor, the gas covered and liquid covered areas of the membrane can be studied separately. At the gas liquid interface, i.e., the area above the liquid covered area of the membrane, liquid species evaporate and are thereupon transported diffusively through the membrane. Also, feed gas and evaporated liquid is diffusively transported through the gas-covered area of the membrane into the permeate channel if no convective flux is occurring.

On the other hand, the sweep gas diffuses via the membrane into the feed channel, but is then transported back by the convection flow. It is important to mention that the overall diffusion through the gas-covered membrane area is strongly affected by the convection flow. To approximate the rate of diffusion of transport through the membrane in the deployed micro contactor without convection flow or liquid feed, a simple calculation based on Fick's law [25] e.g., for the sweep gas (synthetic air), can be used:

$$\dot{V}_{air,diff} = A_p \cdot \Delta y_{air} \cdot \beta_{mem} \tag{14}$$

where $\dot{V}_{air,\,diff}$ is the diffusion flux of air through the membrane, A_p is the cumulative surface area of the pores, Δy_{air} is the gradient of the molar fraction of air through the membrane, and β_{mem} is the mass transfer coefficient in the membrane. Normally, the examined membranes are built with a backer material. In this case, a combined mass transfer coefficient can be calculated as follows:

$$\beta_{mem} = \frac{1}{\frac{1}{\beta_{air,CO_2,backer}} + \frac{1}{\beta_{air,CO_2,active}}} \tag{15}$$

with

$$\beta_{air,CO_2,i} = \frac{D_{air,CO_2,i}}{l_i} \tag{16}$$

where $\beta_{air,CO_2,i}$ is the mass transfer coefficient in the respective layer i, l_i is the thickness of the respective layer i, and $D_{air,CO_2,i}$ is the diffusion coefficient of air in CO_2 in the respective layer i.

2.2.3. Dissolving of Feed Gas

At the gas-liquid interface in the feed stream, gas is dissolved into the liquid. Small channel dimensions generate a high surface to volume ratio, and the diffusion process is enhanced by the small distances. Saturation of the liquid stream with feed gas can therefore be assumed. In the case of carbon dioxide as the feed gas and water as the feed liquid, the maximum solubility of CO_2 can be determined approximately using an equation developed by R. Weiss [27]:

$$k_{CO_2} = \exp\left\{X_1 + X_2 \cdot \frac{100}{T} + X_3 \ln\left(\frac{T}{100}\right)\right\} \cdot \left[\frac{mol}{L \cdot atm}\right] \tag{17}$$

with

$$X_1 = -58.9031; \quad X_2 = 90.5069\frac{1}{K}; \quad X_3 = 22.294\frac{1}{\ln\left(\frac{1}{K}\right)} \tag{18}$$

where k_{CO_2} is the solubility coefficient of CO_2 calculated applying constants X_1, X_2, and X_3 and system temperature T, and p is the pressure in the system.

In addition, as shown by Schüler et al. [28] the solubility of CO_2 in a water-methanol mixture up to 2 mol/L at room temperature is approximately 20% (f(T) = 1.2) higher, at 40 °C nearly constant (f(T) = 1.0), and at 60 °C significantly below 70% (f(T) = 0.3) than in pure water. The maximal-dissolved volume flux of CO_2 $\dot{V}_{CO_2,sol}$ can therefore be calculated.

$$\dot{V}_{CO_2,sol} = f(T) \cdot \dot{V}_1 \cdot k_{CO_2}(T) \cdot p \cdot V_{m,CO_2} \qquad (19)$$

where $\dot{V}_1$ is the volume flux of liquid, f(T) is the solubility multiplier needed for aqueous methanol solutions, and V_{m,CO_2} is the molar volume of the CO_2 gas.

2.2.4. Separation Performance

For the characterisation of the separation efficiency, the separation performance (S_P) is commonly used as an evaluation criterion.

$$S_{P,i} = \frac{\dot{n}_{i,perm}}{\dot{n}_{i,feed,in}} \qquad (20)$$

For a given maximum separable feed gas amount calculated by Darcy's law, the $S_{P,i}$ value is equal to 1, as long as the feed gas inlet volume flow is lower than the maximum separable feed gas amount. If the feed gas inlet volume flow is higher than the highest separable feed gas amount, the $S_{P,i}$ curve shows theoretically a regressive profile, as illustrated in Figure 3.

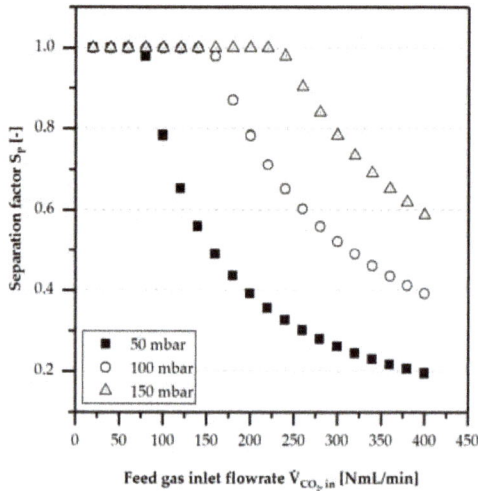

Figure 3. Theoretical $S_{P,i}$ trend according to a maximum separable gas feed amount of 78.3 mL/min, 156.7 mL/min, and 235.0 mL/min calculated by Darcy's law with $\kappa = 7.8 \cdot 10^{-15}$ m, A = 114.5 mm, Δp_{TM} = 50 mbar, 100 mbar, and 150 mbar, η = 17.1 μ Pa s, and l = 200 μm.

2.2.5. Mass Balance

For the characterisation of the separation performance, as well as the measurement of the loss of water/methanol by evaporation, gas saturation and diffusion through the membrane in this work, the composition of the gas from the sweep side outlet of the micro contactor was determined by Fourier-transform infrared spectral analysis (FTIR). The sweep gas (spectrally-inactive synthetic air) was used as a dilution and reference medium. Because of the lack of any chemical reaction, the amount of material is constant and the mass balance of the membrane based micro contactor, as illustrated in Figure 4, can be formulated as:

$$\dot{n}_{feed,in} + \dot{n}_{sweep,in} = \dot{n}_{retentate,out} + \dot{n}_{sweep,out} \tag{21}$$

where $\dot{n}_{feed,in}$ is the feed flux of liquid (aqueous methanol solution) and gas (CO_2), $\dot{n}_{sweep,in}$ is the flux of sweep gas (synthetic air), $\dot{n}_{retentate,out}$ is the flux of gas and or liquid leaving the feed channel and $\dot{n}_{sweep,out}$ is the cumulative flux of sweep gas and permeate leaving the sweep channel.

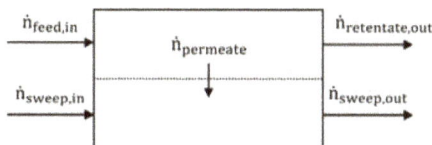

Figure 4. Mass balance model of the separation process.

The cumulative flux of sweep gas and permeate can be calculated using Equation (22)

$$\dot{n}_{sweep,out} = \dot{n}_{sweep,in} + \dot{n}_{perm} \tag{22}$$

if the separation performance S_P is

$$S_P = 1 \tag{23}$$

or

$$S_P < 1 \text{ and } \Delta p_{TM} \geq \Delta p_{TM,crit} \tag{24}$$

It should be noted, that Equation (21) is only valid when no sweep gas is lost through the feed outlet. If the gas separation is complete ($S_{P,i} = 1$), only liquid is leaving the feed channel. At a state of incomplete gas separation, sweep gas can be transported diffusely across the membrane into the feed channel and leave the feed side outlet. In this case, as shown in Section 4.2, a critical transmembrane pressure difference is sufficient to prevent significant sweep gas loss. When these conditions are met, the amount of separated gaseous substances i can be calculated with Equation (25).

$$\dot{n}_{i,perm} = Y_{i,sweep,out} \cdot \dot{n}_{sweep,in} \tag{25}$$

where $Y_{i,sweep,out}$ is the loading of species i in the gas stream leaving the sweep gas channel. Equation (25) can also be written as:

$$\dot{n}_{i,perm} = \frac{Y_{i,sweep,out}}{1 - \sum Y_{i,sweep,out}} \cdot \dot{n}_{sweep,in} \tag{26}$$

with

$$Y_{i,sweep,out} = \frac{\dot{n}_{i,perm}}{\dot{n}_{sweep,out}} \tag{27}$$

where $y_{i,sweep,out}$ is the molar fraction of species i in the gas stream leaving the sweep gas channel measured by calibrated Fourier-transform infrared spectral analysis.

3. Materials and Methods (Experimental)

3.1. Micro Contactor

The experimental investigation of the gas permeability, diffusion, and active-pressure-driven gas/liquid phase separation was performed using the membrane-based micro contactor shown in Figure 5.

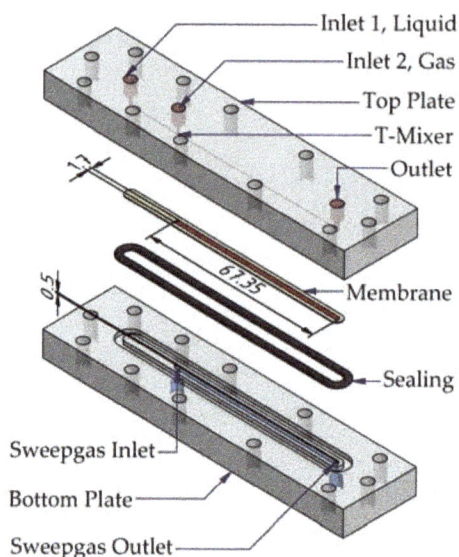

Figure 5. Exploded-view of the membrane based micro contactor with integrated T-junction for two-phase flow generation.

The main components of the membrane-based micro contactor are the bottom plate, the top plate, the milli/micro channel structure, the integrated T-Mixer, the sealing, and the membrane. Both plates have one straight milli/micro channel with a width of 1.7 mm, a height of 0.5 mm, and a length of 67.35 mm. The overall available membrane area for the separation process is 114.5 mm². For a visual observation of the gas/liquid phase separation, the membrane-based micro contactor is made of PMMA, with a contact angle of 60° according to the literature [29].

Two inlets for the feed, i.e., gas and liquid, and one outlet for the retentate are integrated into the top plate. The bottom plate possesses only one inlet and outlet for the sweep gas. In addition, the permeate leaves the micro contactor via the sweep gas outlet. If no sweep gas is used, the inlet of the bottom plate can instead be used as an outlet for the permeate. To generate a two-phase flow, gas and liquid are mixed by a T-junction with a milli/micro channel depth of 0.5 mm, width of 1 mm, and length of 20 mm.

3.2. Porous Membranes

Based on theoretical considerations (Equations (1) and (2)), only hydrophobic membranes for gas/liquid phase separation were chosen as a separation membrane in a micro contactor. Due to the high gas permeability and low pressure loss of porous membranes, the required membrane area for the separation task can be customised to the available pressure gradient, which is the driving force for the gas/liquid phase separation process. Hereby, a compact μDMFC with LOC design and well-defined dimensions for the separation system can be realised.

In this article, four commercial-available porous hydrophobic membranes from two different suppliers (Clarcor Industrial Air and Pall Corporation, Overland Park, KS, USA) were compared. All membranes were built with a two-layer structure including the functional membrane layer and stabilising support layer.

As shown in Table 2, different functional layers and different supporting materials were taken into account for the experimental study. Water entry pressure ranges between 1 to 4.5 bar, as measured by the supplier, which are high and safe enough for use as a separation layer in a μDMFC with LOC design. Furthermore, all membranes are supposed to be temperature resistant in the working range of a μDMFC. Images of the membrane surfaces with a scanning electron microscope (SEM) can be found in Section 4 "Results".

Table 2. Data summary of the polymer based porous hydrophobic membranes. Data extracted from the data sheet provided by Clarcor Air for Aspire® QP955 and Aspire® QL217 and by Pall for Supor® 200PR and Versapor®. Contact angle of water determined by contact angle measurements at lab conditions against air with a measuring accuracy of ±2.5°.

Hydrophobic Membranes	Functional Layer Material	Support Material	Thick-Ness	Pore Size	Contact Angle	Water Entry Pressure
Aspire® QP955 [30]	PTFE	Polyester	200 μm	100 nm	120°	≥4.5 bar
Aspire® QL217 [31]	PTFE	Polypropylen	200 μm	200 nm	120°	≥1.0 bar
Supor® 200PR [32]	Polyethersulfon	Polyester	170 μm	200 nm	138°	≥1.38 bar
Versapor® 200PR [33]	Acrylic Copolymer	Nylon	230 μm	200 nm	120°	≥1.79 bar

3.3. Experimental Setup

3.3.1. Equipment

In Figure 6, a sketch of the experimental setup is shown. It can be divided into fluid dosing ①–④, micro contactor ⑤, pressure regulation ⑥, and analysis ⑦–⑧.

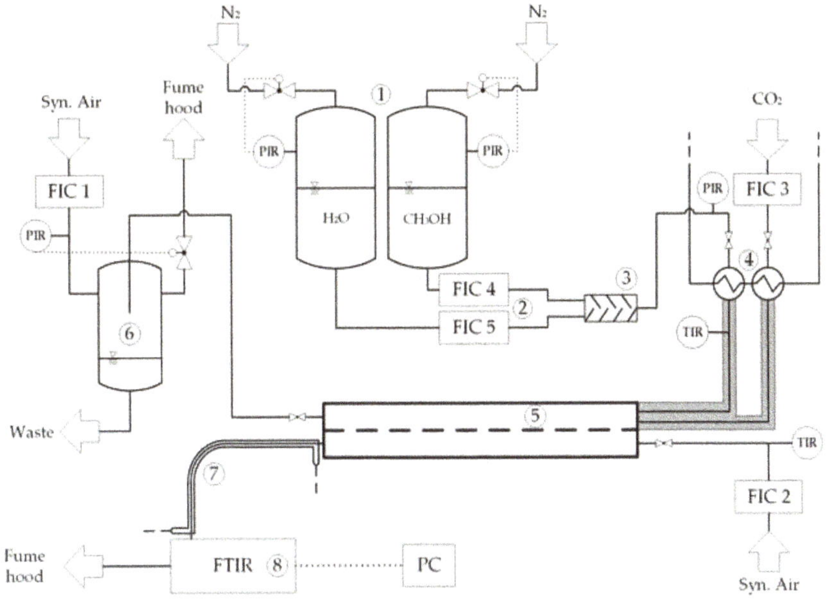

Figure 6. Scheme of the experimental setup.

To ensure fluctuation-free liquid admission, the methanol and water vessels ① were pressurised by a controlled influx of nitrogen. The liquid flow was regulated with mini CORI-FLOW™s, utilising Coriolis force as a measurement principle, whereas gas flows were regulated with an EL-FLOW® thermal mass flow controller (MFC). Both regulation units were manufactured by Bronkhorst High-Tech B.V. The specifications of the employed mini CORI-FLOW™s and MFCs can be found in the supplementary information.

In addition, the liquids were merged in a micro mixer (Part No. 1446-A-9.0) which was designed and manufactured by the Institute of Micro Process Engineering (IMVT, Eggenstein-Leopoldshafen, Germany) ③. Two identically-manufactured cross-flow micro heat exchangers (Part No. 1469-A-1.2

and Part No. 1343-K-1.2), also designed by IMVT, were used to regulate the temperature of the liquid and gas ④. As a thermostating fluid, water was heated and pumped through the micro heat exchangers by a PROLINE P12 thermostat (Lauda Dr. R. Wobser GmbH & Co. KG, Lauda-Königshofen, Germany). Data sheets of the micro mixer and micro heat exchanger can be found in the supplementary information.

Feed gas, sweep gas, and liquid then entered the micro contactor module ⑤ where the separation process took place. At a state of incomplete separation, a two-phase flow streamed out at the outlet of the top plate. For pressure control of the single and two-phase flow, a pressure buffer was integrated ⑥. The temperature of permeate and sweep gas leaving the micro contactor ⑦ were held at a constant 80 °C by a double jacket heat exchanger ⑦. Hereby, the condensation of evaporated liquid species was prevented. Subsequently, the permeate/sweep gas mixture was spectrally analysed in a Fourier transform infrared spectrometer (FTIR) Vector 22 MIR ⑧ (Bruker® Corporation, Billerica, MA, USA) equipped with a 17 mL gas cell with optical windows made from zinc selenide. The multicomponent calibration (CO_2, water, and methanol) of the gas analysis was carried out using the software OPUS 6.5. With a backward calculation using Equation (26), a constant $\dot{n}_{sweep,in}$ and $y_{i,sweep,out}$ from the spectral analysis, the amount of the permeate was determined. For the sweep gas loss experiments or measurements of the retentate, an additional bubble flow meter Definer 220 (Mesa Labs, Lakewood, CO, USA) was used.

The experimental data acquisition and process automation as well as control was done using LabManager®/LabVision® 2.11 (Hitec Zang GmbH, Herzogenrath, Germany) on a standard PC operated with Windows® 7. Verification of the methanol concentration in the water-methanol mixture produced inline at the experimental setup was done using density-based measurements pycnometer at 20 °C (BRAND GMBH & CO KG, Wertheim, Germany). Water contact angle measurements were done manually with an optical contact angle measurement system OCA5 (DataPhysics Instruments GmbH, Filderstadt, Germany). Surface images of the unused and used porous membrane were taken with a JSM-6300 (tungsten cathode) scanning electron microscope (JOEL GmbH Germany, Freising, Germany).

3.3.2. Chemicals

All chemicals were used as bought. Methanol was bought from Merck in p.A. grade. Water was purified with a Milli-Q Reference by Merck® Millipore (serial number F5PA67202D). The electrical resistance of the purified water was measured to be 18.2 MΩ cm. All gases were purchased from Air Liquid in grad N4.5 (separation experiments).

3.4. Operation Conditions

Generally, experiments were performed at a temperature of 20 °C ± 2 °C and a relative humidity of 65% ± 4%. Experiments on separation performance and liquid loss by diffusion and evaporation were measured at feed entry temperatures of 40 °C and 60 °C.

For the orientation tests, a feed entry temperature of 40 °C was chosen. For the determination of the methanol loss, the methanol concentration c_{CH_3OH} of the water-methanol mixture was set to 1 and 2 mol/L, i.e., the usual fuel concentration in an active μDMFC system. For all other experiments, the methanol concentration was generally set to 1 mol/L to study the influence it had on the separation process and membrane material resistance.

3.4.1. Gas Permeability

The gas permeability was measured with a dead-end method for all porous membranes. With this method, the incoming gas volume flow can only leave the membrane micro contactor across the porous membrane. For the determination of the gas permeability, the pressure drop for different gas volumetric flows up to $\dot{V}_{g,feed} = 1000$ NmL/min was detected. Furthermore, the gas permeability and the sweep gas loss for $\Delta p_{TM} = 50, 75,$ and 100 mbar was checked in detail with an open-end method with a constant feed gas volumetric flow $\dot{V}_{g,feed} = 500$ NmL/min and different sweep gas volumetric flows up to $\dot{V}_{g,sweep} = 200$ NmL/min in the permeate channel.

3.4.2. Separation Efficiency

The separation efficiency or performance was evaluated by the following three different criteria:

- Separation performance
- Liquid loss by evaporation and diffusion
- Orientation independence

All experimental parameters for each evaluation criteria are summarised in Table 3.

Table 3. Summary of the experimental parameter for each evaluation criteria.

Evaluation Criteria	c_{CH_3OH} [$\frac{mol}{l}$]	$\dot{V}_{l,feed,in}$ [$\frac{mL}{min}$]	$\dot{V}_{g,feed,in}$ [$\frac{NmL}{min}$]	$\dot{V}_{g,sweep}$ [$\frac{NmL}{min}$]	ϑ [°C]	Δp_{TM} [mbar]
Separation Performance	1	5	50 ... 400	200	20, 40, 60	100
Liquid Loss	0, 1, 2	5	50 ... 400	200	20, 40, 60	100
Orientation Independence	1	5	200, 350	200	40	100

For all porous membranes and evaluation criteria, the transmembrane pressure Δp_{TM} was set to 100 mbar and the liquid volume flow $\dot{V}_{l,feed,in}$ to 5 mL/min. The feed gas volume flow $\dot{V}_{g,feed,in}$ was varied between 50 NmL/min and 400 NmL/min. Additionally, the gas permeability was investigated at separations factors below $S_{P,i} < 1$, indicating the maximum gas permeability of the porous membrane under two-phase flow conditions. Furthermore, the temperature was varied and the methanol concentration c_{CH_3OH} of the water-methanol was set to 1 mol/L.

For the diffusive liquid loss at single phase flow, the Aspire® QL217 membrane was used as an example to detect the real liquid loss by diffusion of water and methanol at room temperature for different methanol concentrations and sweep gas volume flows.

The determination of the orientation independent gas/liquid phase separation with the porous membrane-based micro contactor was checked for four different orientations. The orientation was clockwise radially rotated by 90°, 180°, and 270° with respect to gravity. With these test scenarios, all occurring critical orientations were tested.

3.4.3. Active Membrane Area

Considering the two-phase flow, the membrane area was not completely used during the separation process, especially at full separation. If $S_{P,i} < 1$, a two-phase flow would be still present at the feed outlet. The determination of the active membrane area or participating membrane area during the separation process was done for the Aspire® QL 217 membrane at room temperature with different methanol concentrations. To quantify the active membrane area, the permeability values for single phase and two-phase flow were experimentally determined and afterwards compared. The sweep gas was set to 100 NmL/min.

4. Results

4.1. Gas Permeability

As shown in Figure 7, the transmembrane pressure rose with an increasing feed gas volume flux at room temperature for all membranes.

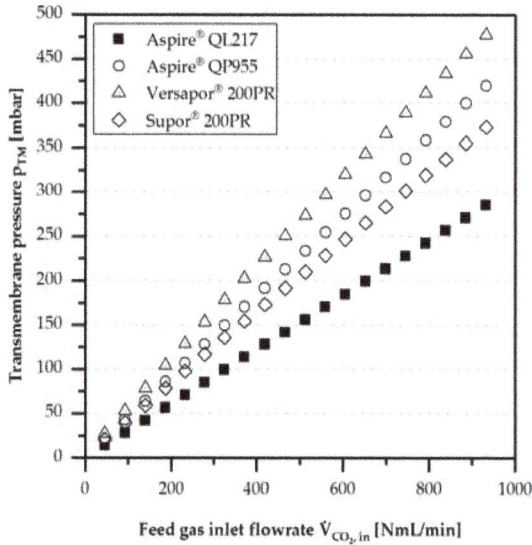

Figure 7. Transmembrane pressure at different $\dot{V}_{CO2}$ for all membranes at $\vartheta = 20\,°C$ and calculated average permeabilities for all membranes at $\vartheta = 20\,°C$.

Gas permeability κ values shown in Table 4 were calculated by rearranging Equation (8). The total membrane thickness (active membrane layer + supporting material) was used for the calculation of the permeability.

Table 4. Gas permeability values for single phase conditions.

Membrane	Gas Permeability κ_i $[10^{-15}\ m^2]$ at 20 °C
Aspire® QL217	9.3
Aspire® QP955	14.0
Versapor® 200PR	10.3
Supor® 200PR	7.9

The highest gas permeability was detected for the membrane Aspire® QL217. The Versapor® 200PR membrane showed the lowest gas permeability, due to the increased thickness of the active membrane layer compared to the other membranes. The relatively low gas permeability of the membrane Aspire® QP955 was caused by the smaller pores (100 nm) compared to the other membranes (200 nm).

4.2. Diffusion

The gas diffusion process from the sweep gas channel into the feed gas channel and vice versa played a significant role at low pressure gradients and low convective flux rates. This can be seen by the exemplary calculated diffusion values for a 10 μm thick active membrane layer with a pore diameter of 200 nm supported by a porous backer material with a thickness of 190 μm. The Knudsen number for air in the active microporous membrane layer is calculated to be Kn = 0.34 with a mean path length $\lambda_{air} = 68$ nm at 1 bar and 298.15 K [34]. Thus, in this case, the diffusion mechanism is a mixture of Knudsen and free molecular diffusion.

However, as a simplification, only pure Knudsen diffusion in the active membrane layer $\left(D_{air,CO_2,active} = D_{air,Kn}\right)$ is assumed to be equal to the highest possible mass transfer coefficient. The Knudsen diffusion value is $D_{air,Kn} = 27.6\ mm^2/s$, whereas the free diffusion value is

$D_{air,CO_2,mol} = 47.5 \text{ mm}^2/\text{s}$. In the real case, it is evident that the diffusion value of the active membrane $D_{air,CO_2,active}$ lies in between $D_{air,Kn}$ and $D_{air,CO_2,mol}$.

In direct comparison, the transport coefficient in the membrane layer $\beta_{air,CO_2,active} = 2.76 \text{ m/s}$ is over 10 times higher than the transport coefficient $\beta_{air,CO_2,backer} = 0.25 \text{ m/s}$ in the thick backer material. It is clear that the backer material limits the mass transfer by diffusion. Due to this, a uniform and free molecular diffusion through the membrane can be assumed, and the Knudsen diffusion can be neglected.

With an average concentration gradient of $\Delta y = 0.5$, considering a pore area of $A_p = A_{mem} 0.5 \approx$ 50 mm, the diffusive volume flux of sweep gas (air) is $V_{air,diff} \approx 350 \text{ NmL/min}$. Additionally, an equimolar flow of CO_2 can be supposed to be transported diffusively in the opposite direction.

It is important to note that this is only a rough estimation. Many aspects, including the diffusive mass transport within feed and sweep channel and exact membrane characteristics such as tortuosity, porosity, and pore length, are not considered. Nevertheless, it can be assumed that the polymer membranes used is this article, just like those considered in the model membrane, have a very low mass transfer resistance.

In addition, the sweep gas loss by diffusion was detected at different sweep gas fluxes and transmembrane pressure gradients, as shown in Figure 8. A clear correlation between low pressure gradients and high concentration gradient-driven sweep gas loss by diffusion was observed.

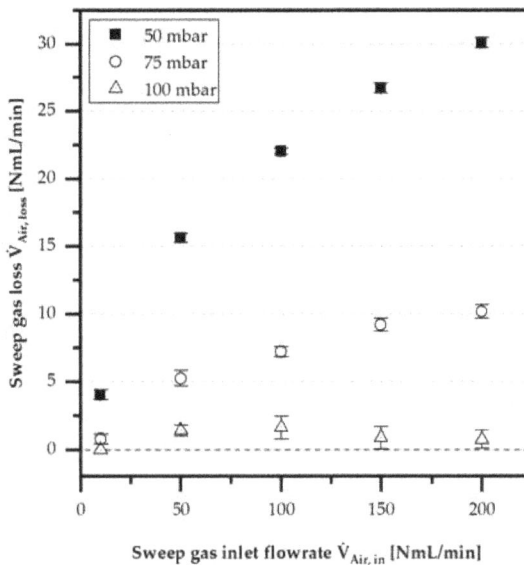

Figure 8. Measured sweep gas loss through a Aspire® QL217 membrane for different sweep gas flows at $\vartheta = 20°C$ and $\dot{V}_{gas,feed,in} = 500 \text{ NmL/min}$.

At a pressure gradient of 100 mbar, these losses became insignificant. Diffusive transport of the feed gas also inevitably followed this behaviour. It can be concluded that diffusive mass transfer phenomena will be irrelevant at pressure gradients higher or equal to $\Delta p_{TM,crit} = 100 \text{ mbar}$.

4.3. Separation Efficiency

The effectiveness of a separation process was assessed by the separation performance, liquid loss, and orientation independence. A detailed overview of the results and a discussion are provided below.

4.3.1. Separation Performance

In principle, the porous membrane-based micro contactor can be installed downstream of the μDMFC system to achieve a flat, compact, and portable LOC design. By doing so, an open or closed breather can be realised. In the case of the open breather, the membrane is visible from the outside, and directly separates the CO_2 into the environment, whereas the closed breather has an additional channel for flushing by a sweep gas. However, in both cases, isothermal operating conditions cannot be assumed due to the transient operating mode of the portable μDMFC.

In all experiments where the separation efficiency was investigated at increased fluid inlet temperatures and without isothermal micro contactor operating conditions, liquid condensation under single- and two-phase conditions at the membrane support, i.e., the sweep gas channel, was observed. First, the liquid species evaporated at the membrane side into the gas phase. Thereafter, the vapour was transported by diffusion or convection through the active and supporting layer of the membrane. Subsequently, the vapour condensed at the membrane support due to the lower temperature within the sweep gas channel as shown in Figure 9.

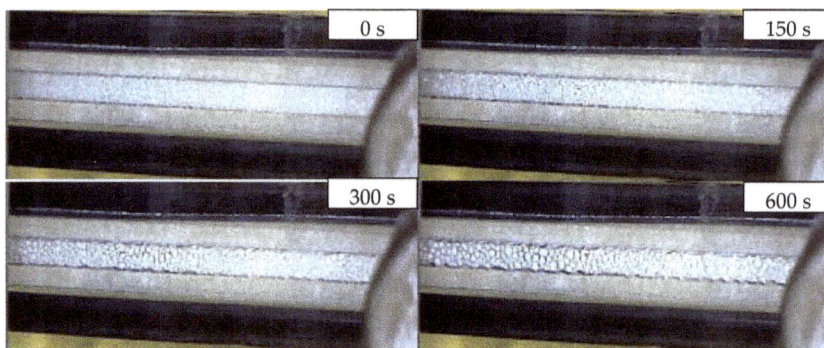

Figure 9. Degree of condensation of the evaporated and diffusively transported liquid species at 40 °C within the sweep gas channel on the supporting material, observed for different operating times. Illustrated pictures are for following operating times 0 s, 150 s, 300 s, and 600 s using $\dot{V}_{l,feed,in} = 5$ mL/min, $\dot{V}_{g,feed,in} = 0$ NmL/min, and $\dot{V}_{g,sweep,in} = 0$ NmL/min.

As a consequence, the pores of the porous membrane got blocked and the separation performance decreased rapidly after 10 min. To reduce the diffusive flux (part of the liquid loss of water and methanol) and to avoid the condensation, a micro heat exchanger installed downstream after the μDMFC should be taken into account for the μDMFC with LOC design.

Another possibility for avoiding condensation is an increased sweep gas flow up to 200 NmL/min. Hereby, an increased liquid loss of water and methanol by liquid species diffusion during the experiments without condensation was observed (see Section 4.3.2.). Moreover, the constantly supplied sweep gas (synthetic air) was used as reference and dilution to detect inline the real separated CO_2 amount by FTIR measurements. The experimental results for $\vartheta_{inlet} = 20$ °C, 40 °C, and 60 °C are shown in Figure 10.

All porous membranes showed complete separation behaviour ($S_P \geq 1$ or $S_P =$ const.) for CO_2 gas feed volume flows between 100–150 NmL/min. A decrease of S_P was observed at CO_2 feed gas volume flows above 100–150 NmL/min. In both cases, a dependence on the inlet temperature ϑ_{inlet} was noticed due to the liquid evaporation increasing the amount of gas to be separated, as well as decreasing surface tension and viscosity of the water-methanol mixture. Nevertheless, the curve characteristics for all porous membranes were almost identical, except for the maximum separation amount.

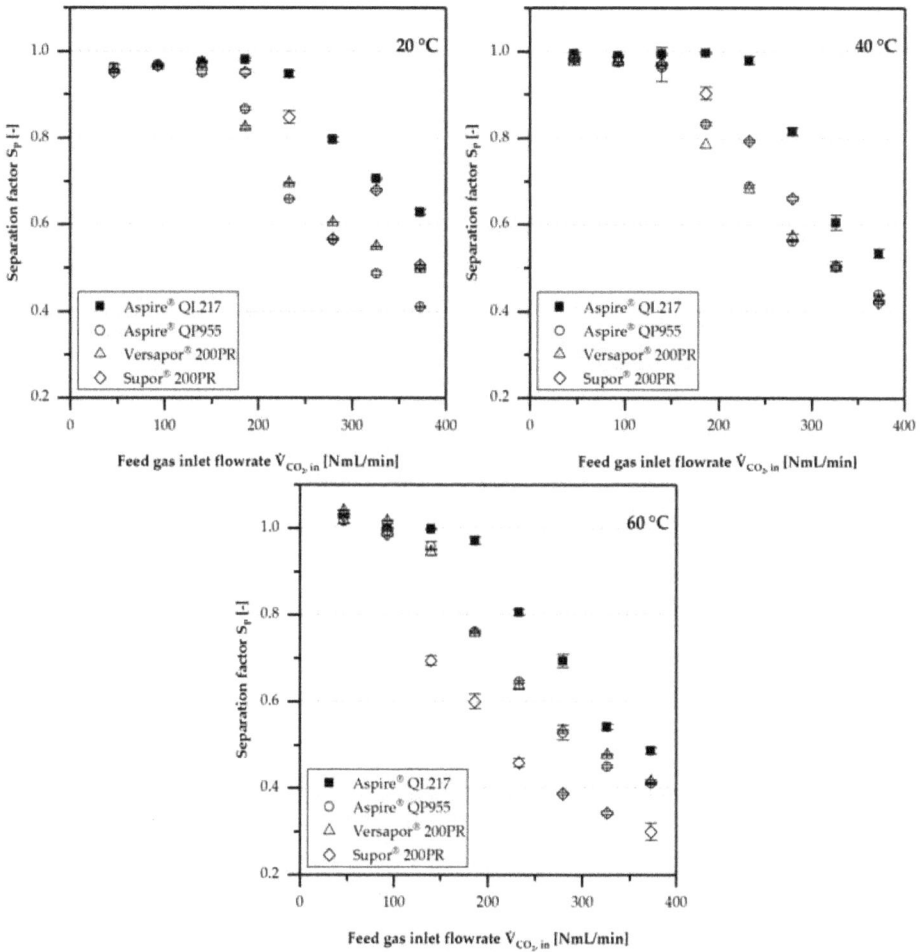

Figure 10. Comparison of the separation performance S_{P,CO_2} for different $\dot{V}_{g,feed,in}$ at $\vartheta_{inlet} = 20\,°C$, $40\,°C$, and $60\,°C$ ($\dot{V}_{l,feed,in} = 5\,mL/min$, $\Delta p_{TM} = 100\,mbar$, and $c_{CH_3OH} = 1\,mol/L$.

The turning point of S_P indicated the maximum separable amount of CO_2 at any given transmembrane pressure as described by Darcy's law. From this point, a characteristic regressive curve was detected, agreeing with the theory.

At room temperature, the S_P values for a complete separation were observed to be slightly below 1, whereas at $60\,°C$, the S_P values were slightly above 1. In both cases, the gas solution and dissolution processes of CO_2 into the water-methanol mixture during the separation were responsible for this phenomenon.

Moreover, only the porous membrane Supor® 200PR showed a dramatic decrease in S_P at $\vartheta_{inlet} = 60\,°C$. After decreasing the temperature back to $\vartheta_{inlet} = 40\,°C$, nearly the same S_P values were detected. It is assumed that a reversible temperature and methanol-induced change in the porous membrane took place, leading to reduced gas permeability. Several repetitions with increasing and decreasing inlet temperatures showed the same behaviour.

After the experiments, the membrane surfaces of all used membranes were investigated by SEM and compared with the unused ones, as shown in Figure 11. For the PTFE-based membranes Aspire®

QP955 and Aspire® QP217, no significant changes on the membrane surface could be observed. The Supor® 200PR membrane instead showed an irreversible closing effect of the pores by chemical or thermal bonding of the membrane material (red squares). In the case of the porous membrane Versapor® 200PR, an increase in surface porosity by visually larger pore diameter was noticed.

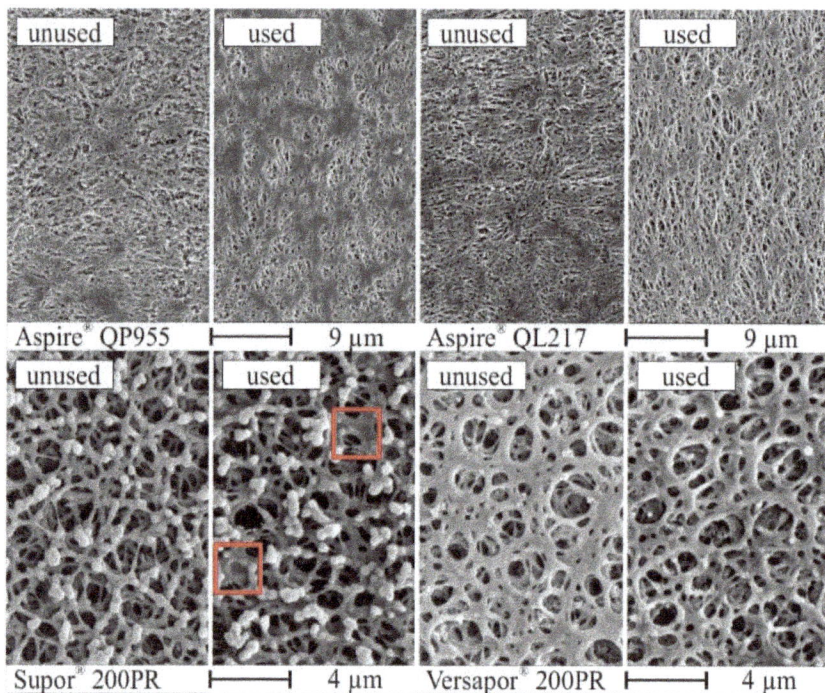

Figure 11. Images of the membrane surface by scanning electron microscopy after usage as separation layer for all experimental investigations.

During short term tests using Supor® 200PR and Versapor® 200PR membranes, the detected changes of the surface structure did not affect the separation performance. Nevertheless, it was concluded that the acrylic copolymer-based membrane (Versapor 200PR) was degraded at the top of the surface by the water-methanol mixture. For the observed surface behaviour of the polyethersulfon-based membrane (Supor® 200PR), additional investigations are needed.

The comparison of the gas permeability under single and two-phase flow conditions in Table 5 showed that the gas permeability of the pure gas (Section 4.1) was significantly higher than with two-phase flow conditions. It is assumed that this result is in general due to the not fully utilised membrane area during the separation process.

Table 5. Comparison of the gas permeability under single and two-phase flow conditions.

Membrane	κ_i [10^{-15} m^2] at 20 °C			A_i [mm^2]
	κ_g	κ_{TPF}	$\frac{\kappa_{TPF}}{\kappa_g}$	A_{active}
Aspire® QP955	9.3	7.2	0.77	88.7
Aspire® QL217	14.0	10.6	0.76	86.7
Supor® 200PR	10.3	9.0	0.87	100.0
Versapor® 200PR	9.1	7.7	0.85	96.9

However, the liquid part of the two-phase flow partially reduced the active membrane area at slug, transient, and corner flow state. In general, the membrane was covered by the liquid slugs at slug flow conditions, whereas at under liquid corner flow conditions, i.e., at high feed gas volumetric flows, only the edges were covered by the liquid. In addition, a proportion of the pores can be physically blocked by the liquid. A detailed investigation of the utilised membrane area can be found in Section 4.3 "Active Membrane Area".

For all CO_2 gas feed volume flows, the Aspire® QL217 from CLARCOR Industrial Air, followed by the Supor 200PR from Pall delivered the best separation performance and gas permeability values for 20 °C and 40 °C.

4.3.2. Liquid Loss of Water/Methanol

Knowledge about liquid loss of water and methanol by evaporation and diffusion in a porous membrane-based micro contactor operating as a closed or open breather for a µDMFC with LOC design is an important evaluation criterion. Through a multi-variant calibration, all species in the sweep gas outlet leaving the micro contactor could be determined. Thus, the real amount of liquid loss during the separation process was measurable.

In order to exclude condensation in the sweep gas channel and piping, a higher sweep gas volume flow was set. Hereby, the proportion of liquid loss by diffusion was higher than at lower sweep gas volume flows, as shown in Figure 12, for a porous membrane-based micro contactor without TPF.

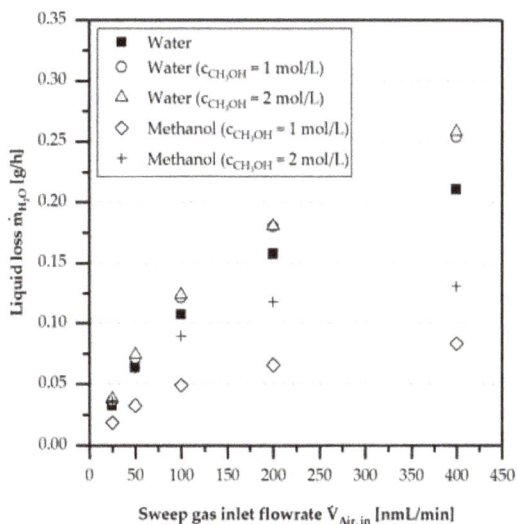

Figure 12. Diffusive liquid loss for water and methanol under single phase flow conditions (liquid) of the porous membrane depending on the sweep gas volume flow for Aspire® QL217 at 20 °C and $\dot{V}_{l,feed,in} = 5$ mL/min.

Furthermore, the liquid loss of water was larger than the liquid loss of methanol. For the liquid loss of methanol, a proportional behaviour to the methanol concentration was observed. Simultaneously, increased liquid loss of water was detected due to a higher vapour pressure of the water-methanol mixture. In addition, a slight contribution to the liquid loss by a decreasing surface tension and viscosity is assumed.

In direct comparison, the liquid loss by diffusion was significantly higher than the liquid loss through feed gas saturation by evaporation. Furthermore, it can be clearly seen in Figures 13 and 14 that the liquid loss of water and methanol increased with rising temperatures and feed gas volume flows. Additionally, the impact of increasing volume flows was lower than the increasing temperatures on the liquid loss of water and methanol.

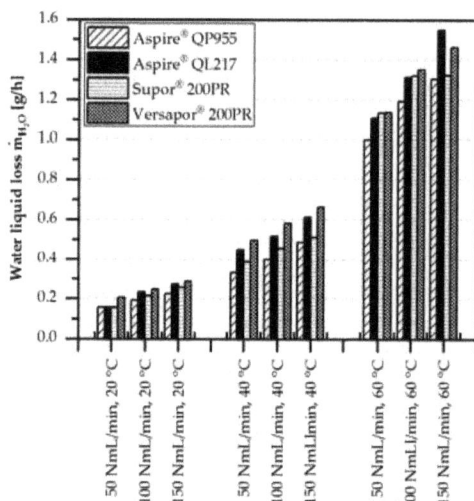

Figure 13. Overall liquid loss for water and for all porous membrane under two-phase flow conditions for different feed inlet temperatures ($\vartheta_{inlet} = 20\ °C$, $40\ °C$, and $60\ °C$) and feed gas volume flows $\left(\dot{V}_{l,feed,in} = 5\,\frac{mL}{min}\ \text{and}\ \dot{V}_{g,feed,in} = 50\,\frac{NmL}{min},\ 100\,\frac{NmL}{min},\ \text{and}\ 150\,\frac{NmL}{min}\right)$.

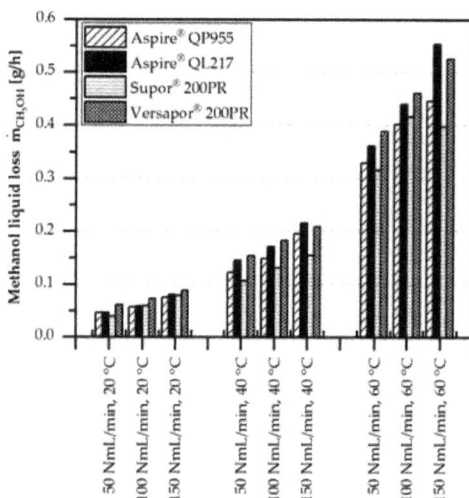

Figure 14. Overall liquid loss for methanol and for all porous membrane under two-phase flow conditions for different feed inlet temperatures ($\vartheta_{inlet} = 20\ °C$, $40\ °C$, and $60\ °C$) and feed gas volume flows $\left(\dot{V}_{l,feed,in} = 5\,\frac{mL}{min}\ \text{and}\ \dot{V}_{g,feed,in} = 50\,\frac{NmL}{min},\ 100\,\frac{NmL}{min},\ \text{and}\ 150\,\frac{NmL}{min}\right)$.

The Aspire® QP955 and Supor® 200PR had the lowest liquid loss for water and methanol over the whole tested range compared to the other porous membranes. In the case of the Aspire® QP955 membrane, the smaller pore diameter ($d_p = 100$ nm) was probably responsible for this behaviour, whereas for the Supor® 200PR, the surface structure of the membrane was responsible. The highest liquid losses of water and methanol were detected for the Aspire® QL217 und Versapor® 200PR.

Based on the presented results, it can be concluded that the liquid loss due to the diffusion was the largest in a closed breather with additional sweep gas. The liquid loss in a closed or open breather without sweep gas is mainly caused by gas saturation through evaporation.

4.3.3. Orientation Independence

The portability of a porous membrane-based micro contactor for the gas/liquid phase separation depends primarily on the orientation independent working principle. Theoretically, as mentioned in Section 2 "Theoretical Background", a calculated Confinement number of Co = 3.5 and the Etövos number of Eo = 0.1 should be sufficient for an orientation-independent gas/liquid phase separation. For the calculation of Co and Eo, a micro channel filled with water + air, a height of 0.5 mm, and a width of 1.7 mm was assumed.

Nevertheless, different orientations, as shown in Figure 15, were tested for all membranes at complete separation state (S_P = 1 or S_P = const.) and incomplete separation state (S_P < 1) were tested, to confirm the orientation independent gas/liquid phase separation. As an example, the result for the Aspire® QL217 is shown in Figure 15. With a feed gas volume flow rate of 200 NmL/min and 350 NmL/min, a $S_P \leq 1$ and $S_P \leq 0.6$ was detected, respectively.

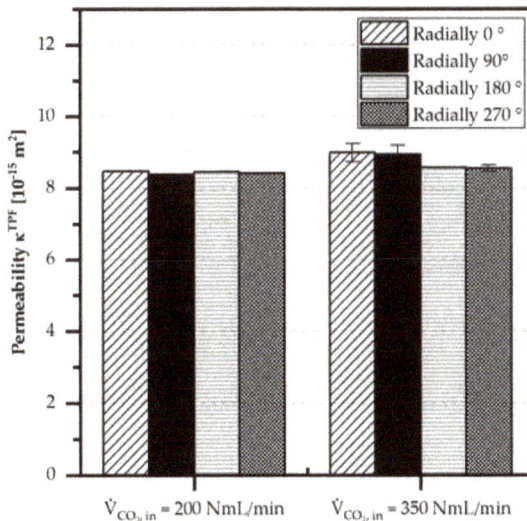

Figure 15. Orientation independence tests for Aspire® QL217 ($\dot{V}_{l,feed,in}$ = 5 mL/min, ϑ_{inlet} = 40 °C, Δp_{TM} = 100 mbar, and c_{CH_3OH} = 1 mol/L).

The orientations had only a minimal influence on the separation performance—less than 3%—for a feed gas volume flow rate of 200 NmL/min. For a feed gas volume flow rate of 350 NmL/min, an influence of less than 5% was observed. Identical results were also obtained for the other membranes. Based on the results shown above, an orientation independent separation performance can be assumed.

4.4. Active Membrane Area

When two-phase flow is present in the feed channel of the micro contactor module, a fraction of the overall available membrane is covered with liquid, and can therefore not be used for gas separation. Only the gas-covered part of the membrane, defined as active membrane area, contributes to the separation process. The active membrane area, calculated using Equation (9) and κ_g from Table 4, grows with increasing feed gas flux.

In Figure 16, this phenomenon is shown for the membrane Aspire® QL217 and for different feed liquid methanol concentrations. At a state of complete gas separation ($S_P = 1$ or $S_P = $ const.), the active membrane area increased linearly with increasing flow rate of the feed gas. When the feed gas flux exceeded the separation capability of the membrane ($S_P < 1$), the calculated values for the active membrane area shows exponentially-decreasing behaviour.

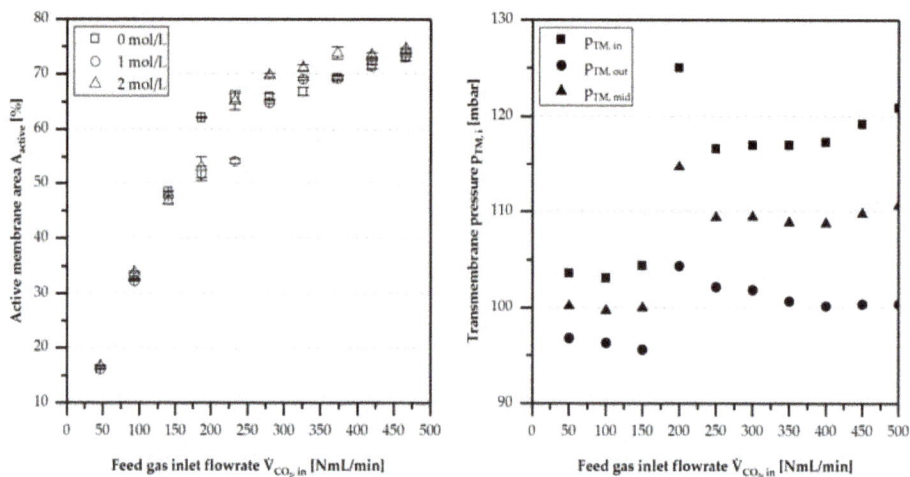

Figure 16. Active membrane area (calculated with Equation (9)) of membrane QL 217 at $\vartheta = 20\,°C$, $\Delta p_{TM} = 100$ mbar, $\dot{V}_{liq,feed,in} = 5$ mL/min, and $\dot{V}_{gas,\,sweep,\,in} = 10$ mL/min for different methanol concentrations and different $\dot{V}_{gas,feed,in}$ (**left image**). Development of the transmembrane pressure gradient with increasing $\dot{V}_{gas,feed,in}$ (**right image**).

In addition, an increased feed gas flux led to a higher pressure gradient, and subsequently, to a higher gas permeation flux. At the same time, the gas/liquid ratio in the feed channel and the mean velocity increased. Through this, the residence time of the liquid on the membrane was reduced by pushing the liquid out of the feed channel more rapidly, resulting in a decreased membrane area blocked by liquid.

The high discontinuity of the measurement results at 200–300 NmL/min was caused by the instability of the separation at the point where the separation first became incomplete and insignificant amounts of gas could leave the feed channel outlet. At this point, pressure increased abruptly because of the additional pressure loss caused by the gas pushing the liquid in the line out, leaving the feed channel. This effect was dynamic, and pressure fluctuation could also be high at this point. The separation stabilised when more feed gas was used.

The concentration of methanol in the feed liquid flow—in the range of 0–2 mol/L—had no direct influence on the available membrane area or the active membrane area during the separation process.

5. Conclusions

In general, the applicability of porous membranes as a separation layer under µDMFC working conditions (physically and chemically) was confirmed. Due to the high capillary pressure in the small pores, the permeation of liquid was prevented, while gas could be transported via the pores of the membrane across the membrane by applying a pressure gradient. In addition, the diffusion flow via these membranes is substantial, and should not be neglected if the separation exceeds the capability of the available membrane area and a high sweep gas flux is used.

In the investigated separation module, the pressure in the gas bubble or phase, which is strongly dependent on the channel size, was negligible compared to the pressure gradient required for an efficient gas separation process. Furthermore, the capillary pressure was always overlain by the pressure drop caused by the active transport of fluids through the narrow channel. Through this observation, it can be concluded that the gas separation solely driven by the capillary pressure in the gas phase located in the channel cannot be realised in a compact and active driven μDMFC.

To increase the capillary pressure in the gas bubble or gas phase, the channel size has to be reduced by an order of 10–20. However, by reducing the channel size, the pressure drop due to the smaller channel would increase dramatically. A parallel arrangement of smaller channels instead of a single channel could be helpful to normalise the pressure drop again, but the distributions of two-phase into the smaller channels is not homogenous enough and less controllable. As a result, uncontrollable short-circuit flows of unseparated gas can occur.

Furthermore, the gas separation was found to follow Darcy's law describing the separated gas amount as proportional to the pressure gradient over the membrane. To obtain complete phase separation under some of the investigated conditions, it was observed that for an effective gas separation, a transmembrane pressure of at least 100 mbar was necessary. In addition, it is known that the low bubble pressure, i.e., the capillary pressure of the gas phase or bubble within the micro channel, amounts to a single digit mbar value, so that the driving force has to be a sufficiently high transmembrane pressure. With an area of 114.5 mm^2, all membranes were capable of separating at least 100 NmL/min CO_2 under μDMFC working conditions using a pressure gradient of 100 mbar. This is sufficient for a typical 20 W DMFC producing 93 NmL/min CO_2 gas at a considered efficiency of 41% [35].

The diffusion of the liquid species and the amount of sweep gas also affected the diffusion of feed gas into the sweep channel and vice versa. The diffusion process was also dependent on the convective gas flow through the membrane, and thereby, was dependent on the applied pressure gradient. At high pressure gradients, diffusion rates of feed and sweep gas were found to be insignificant. It was observed that the convective flux is predominant at an empirically-determined, critical transmembrane pressure gradient $\Delta p_{TM,crit}$ of 100 mbar, suppressing the diffusion flow of the sweep gas completely.

Finally, alternatives to polymer-based membranes should be investigated to overcome negative properties such as swelling or chemical resistance against methanol. At present, metallic microsieves coated with hydrophobic layers are considered as a potential substitute, and are being studied at the Institute of Micro Process Engineering. With respect to the separation system and its integration in a μDMFC, further research is still required. Different channel geometries and channel coatings could provide a better liquid/gas distribution during the separation, and could increase the active membrane area for the gas separation. Finally, using a prototype μDMFC to create the two-phase-flow feed stream to accurately replicate the conditions that would be expected in a complete integrated system, could improve the separation process under transient working conditions.

Supplementary Materials: The following are available online at http://www.mdpi.com/2305-7084/2/4/55/s1.

Author Contributions: Conceptualisation, K.M.D.; methodology, K.M.D.; formal analysis, K.M.D. and V.W.; investigation, V.W. and K.M.D.; resources, KIT; data curation, V.W. and K.M.D., writing—original draft preparation, K.M.D.; writing—review and editing, K.H.-S.; visualisation, K.M. and V.W.; supervision, R.D.; project administration, K.M.D. and K.H.-S.; funding acquisition, K.H.-S. and R.D.

Funding: This work was funded and supported by the BMWi and AiF through IGF-Project 18741 N.

Acknowledgments: The authors acknowledge support of this work by staff of the Institute for Micro Process Engineering at Karlsruhe Institute of Technology. We also thank "Clarcor Industrial Air" and "Pall Corporation" for the graciously provided porous membranes.

Conflicts of Interest: The authors declare no conflict of interest.

Nomenclature

Latin Symbols

A	area (m^2)
Co	confinement number (-)
d	diameter (m)
D	diffusion coefficient ($m^2\ s^{-1}$)
d_h	hydraulic dimeter (m)
Eö	Eötvös number (-)
f	multiplication factor (-)
g	standard gravity (9.981 kg m s^{-2})
h	height (m)
H	mean curvature (m^{-1})
k	curvature of sphere (m^{-1})
k_{CO_2}	solubility coefficient (mol L^{-1} atm^{-1})
K	Parameter (-)
k_b	Boltzmann constant (1.38 10^{-23} J K^{-1})
Kn	Knudsen number (-)
l	thickness (m)
m	mass (kg)
M	molar mass (kg mol^{-1})
$\dot{n}$	molar flux (mol s^{-1})
p	pressure (Pa)
r	radius (m)
R	radius of channel geometry (m)
S_p	separation performance (-)
T	temperature (K)
U	perimeter (m)
V	volume (m^3)
$\dot{V}$	volume flux ($m^3\ s^{-1}$)
V_m	molar volume ($m^3\ mol^{-1}$)
v	diffusion volume (-)
w	width (-)
x/y	molar fraction (-)
X	solubility constant (-)
Y	molar loading (-)

Subscripts

active	active
air	air
backer	backer material
cap	capillary
coll.	collision
crit	critical
CH_3OH	methanol
CO_2	carbon dioxide
diff	diffusive
feed	feed
gas	gas
H_2O	water
in	in
inlet	inlet
Kn	Knudsen

liq	liquid
lg	liquid-gas-interface
mem	membrane
mol	molecular
nw	non-wetted
out	out
p	pores
perm	permeate
sg	solid-gas-interface
sl	solid-liquid-interface
sol	dissolved
sweep	sweep
tm	trans membrane
TPF	two-phase-flow
w	wetted
t,b,l,r	top, bottom, left, right

Greek symbols

β	mass transfer coefficient (m s^{-1})
η	dynamic viscosity (Pa s)
ϑ	temperature (°C)
κ	permeability (m^2)
λ	mean free path length (m)
ρ	density (kg m^{-3})
σ	surface tension (N m^{-1})
σ_{coll}	collision diameter (m)
φ	contact angle (°)
Δ	gradient (-)
ω	mass fraction (-)

Abbreviations

µDMFC	micro-direct methanol fuel cell
DIK	Deutsches Institut für Kautschuktechnologie
FTIR	Fourier-transform infrared
IMVT	Institute for Micro Process Engineering
KIT	Karlsruhe Institute of Technology
LOC	lab on a chip
PMMA	polymethylmethacrylate
PP	polypropylene
PTFE	polytetrafluoroethylene
SEM	scanning electron microscope
S_P	separation performance
TPF	two-phase flow
ZBT	Zentrum für Brennstoffzellentechnik

References

1. Schaevitz, S.B. Powering the wireless world with MEMS. *Proc. SPIE* **2012**, *8248*, 1–15. [CrossRef]
2. Krewer, U. Portable Energiesysteme: Von elektrochemischer Wandlung bis Energy Harvesting. *Chem. Ing. Tech.* **2011**, *83*, 1974–1983. [CrossRef]
3. Rummich, E. Energiespeicher. *Elektrotech. Inftech.* **2013**, *130*, 143–144. [CrossRef]
4. Li, X.; Faghri, A. Review and advances of direct methanol fuel cells (DMFCs) part I: Design, fabrication, and testing with high concentration methanol solutions. *J. Power Sources* **2013**, *226*, 223–240. [CrossRef]
5. Novosolution. Freezing Points of Methanol/Water Solutions. Available online: http://novosolution.ca/images/Freezing-Points-Methanol.pdf (accessed on 20 February 2018).

6.   Paust, N.; Krumbholz, S.; Munt, S.; Müller, C.; Koltay, P.; Zengerle, R.; Ziegler, C. Self-regulating passive fuel supply for small direct methanol fuel cells operating in all orientations. *J. Power Sources* **2009**, *192*, 442–450. [CrossRef]

7.   Zenith, F.; Krewer, U. Modelling, dynamics and control of a portable DMFC system. *J. Process Control* **2010**, *20*, 630–642. [CrossRef]

8.   Zenith, F.; Weinzierl, C.; Krewer, U. Model-based analysis of the feasibility envelope for autonomous operation of a portable direct methanol fuel-cell system. *Chem. Eng. Sci.* **2010**, *65*, 4411–4419. [CrossRef]

9.   Meng, D.D.; Kim, J.; Kim, C.-J. A distributed gas breather for micro direct methanol fuel cell (μ-DMFC). In Proceedings of the Sixteenth Annual International Conference on Micro Electro Mechanical Systems, Kyoto, Japan, 23 January 2003. [CrossRef]

10.   Lee, S.-W.; Wong, S.-C. Design and fabrication of multidirectional microbubble separator. *Proc. SPIE* **2005**, *5718*, 194–199. [CrossRef]

11.   Amon, C.H.; Yao, S.-C.; Tang, X.; Hsieh, C.-C.; Alyousef, Y.; Vladimer, M.; Fedder, G.K. Micro-electro-mechanical systems (MEMS)-based micro-scale direct methanol fuel cell development. *Energy* **2006**, *31*, 636–649. [CrossRef]

12.   Alexander, B.R.; Wang, E.N. Design of a Microbreather for Two-Phase Microchannel Heat Sinks. *Nanoscale Microscale Thermophys. Eng.* **2009**, *13*, 151–164. [CrossRef]

13.   Kraus, M.; Krewer, U. Experimental analysis of the separation efficiency of an orientation independent gas/liquid membrane separator. *Sep. Purif. Technol.* **2011**, *81*, 347–356. [CrossRef]

14.   Meng, D.D.; Kim, J.; Kim, C.-J. A degassing plate with hydrophobic bubble capture and distributed venting for microfluidic devices. *J. Micromech. Microeng.* **2006**, *16*, 419–424. [CrossRef]

15.   Meng, D.D.; Cubaud, T.; Ho, C.-M.; Kim, C.-J. A Methanol-Tolerant Gas-Venting Microchannel for a Microdirect Methanol Fuel Cell. *J. Microelectromech. Syst.* **2007**, *16*, 1403–1410. [CrossRef]

16.   Meng, D.D.; Kim, C.J. An active micro-direct methanol fuel cell with self-circulation of fuel and built-in removal of CO2 bubbles. *J. Power Sources* **2009**, *194*, 445–450. [CrossRef]

17.   Xu, J.; Vaillant, R.; Attinger, D. Use of a porous membrane for gas bubble removal in microfluidic channels: Physical mechanisms and design criteria. *Microfluid. Nanofluid.* **2010**, *9*, 765–772. [CrossRef]

18.   David, M.P.; Steinbrenner, J.E.; Miler, J.; Goodson, K.E. Adiabatic and diabatic two-phase venting flow in a microchannel. *Int. J. Multiph. Flow* **2011**, *37*, 1135–1146. [CrossRef]

19.   Fazeli, A.; Moghaddam, S. Microscale phase separator for selective extraction of CO_2 from methanol solution flow. *J. Power Sources* **2014**, *271*, 160–166. [CrossRef]

20.   Diversified Enterprises. Surface Energy Data for PTFE. 2018. Available online: https://www.accudynetest.com/polymer_surface_data/ptfe.pdf (accessed on 20 February 2018).

21.   Huh, D.; Kuo, C.-H.; Grotberg, J.B.; Takayama, S. Gas-liquid two-phase flow patterns in rectangular polymeric microchannels: Effect of surface wetting properties. *New J. Phys.* **2009**, *11*, 75034. [CrossRef] [PubMed]

22.   Serizawa, A.; Feng, Z.; Kawara, Z. Two-phase flow in microchannels. *Exp. Therm. Fluid Sci.* **2002**, *26*, 703–714. [CrossRef]

23.   Suo, M. Two-Phase Flow in Capillary Tubes. Ph.D. Thesis, Massachusetts Institute of Technology, Cabridge, MA, USA, 1960.

24.   Brauner, N.; Maron, D.M. Identification of the range of 'small diameters' conduits, regarding two-phase flow pattern transitions. *Int. Commun. Heat Mass Transf.* **1992**, *19*, 29–39. [CrossRef]

25.   Cussler, E.L. *Diffusion. Mass Transfer in Fluid Systems*, 3rd ed.; Cambridge University Press: Cambridge, UK, 2011.

26.   Fuller, E.N.; Giddings, J.C. A Comparison of Methods for Predicting Gaseous Diffusion Coefficients. *J. Chromatogr. Sci.* **1965**, *3*, 222–227. [CrossRef]

27.   Weiss, R.F. Carbon dioxide in water and seawater: The solubility of a non-ideal gas. *Mar. Chem.* **1974**, *2*, 203–215. [CrossRef]

28.   Schüler, N.; Hecht, K.; Kraut, M.; Dittmeyer, R. On the Solubility of Carbon Dioxide in Binary Water–Methanol Mixtures. *J. Chem. Eng. Data* **2012**, *57*, 2304–2308. [CrossRef]

29.   Diversified Enterprises. Surface Energy Data for PMMA. 2018. Available online: https://www.accudynetest.com/polymer_surface_data/pmma_polymethylmethacrylate.pdf (accessed on 20 February 2018).

30.   Clarcor Industrial Air. *Data Sheet aspire® ePTFE Membrane Laminate*; Aspire® QP955; Clarcor Industrial Air: Overland Park, KS, USA, 2014.

31.   Clarcor Industrial Air. *Data Sheet aspire® ePTFE Membrane Laminate*; Aspire® QL217; Clarcor Industrial Air: Overland Park, KS, USA, 2014.

32. Pall Corporaton. Data Sheet Supor® R Membrane. Supor 200PR. 2009. Available online: http://www.pall.de/pdfs/misc/IMGSRMEN.pdf (accessed on 20 February 2018).
33. Pall Corporaton. Data Sheet Versapor® R Membrane. Versapor 200PR. 2010. Available online: http://www.pall.de/pdfs/misc/IMGVRMEN.pdf (accessed on 20 February 2018).
34. Jennings, S.G. The mean free path in air. *J. Aerosol Sci.* **1988**, *19*, 159–166. [CrossRef]
35. Rashidi, R.; Dincer, I.; Naterer, G.F.; Berg, P. Performance evaluation of direct methanol fuel cells for portable applications. *J. Power Sources* **2009**, *187*, 509–516. [CrossRef]

chemengineering

MDPI

Article

Solar Energy Assisted Membrane Reactor for Hydrogen Production

Barbara Morico [1], Annarita Salladini [1], Emma Palo [2] and Gaetano Iaquaniello [2,*]

[1] Processi Innovativi srl, Via di Vannina 88, 00156 Rome, Italy; morico.b@processiinnovativi.it (B.M.); salladini.a@processiinnovativi.it (A.S.)

[2] KT—Kinetics Technology S.p.A., Viale Castello della Magliana 27, 00148 Rome, Italy; e.palo@kt-met.it

* Correspondence: g.iaquaniello@kt-met.it; Tel.: +39-06-60216231

Received: 9 November 2018; Accepted: 2 January 2019; Published: 15 January 2019

Abstract: Pd-based membrane reactors are strongly recognized as an effective way to boost H_2 yield and natural gas (NG) conversion at low temperatures, compared to conventional steam reforming plants for hydrogen production, thereby representing a potential solution to reduce the energy penalty of such a process, while keeping the lower CO_2 emissions. On the other hand, the exploitation of solar energy coupled with a membrane steam reformer can further reduce the environmental impact of these systems. On this basis, the paper deals with the design activities and experimentation carried out at a pilot level in an integrated prototype where structured catalysts and Pd-based membranes are arranged together and thermally supported by solar-heated molten salts for steam reforming reaction

Keywords: membrane reactor; Pd-based membrane; hydrogen; steam reforming; solar energy

1. Introduction

Pd-based membrane reactors are strongly recognized as an effective way to boost H_2 yield and natural gas (NG) conversion at low temperatures, compared to conventional steam reforming plants for hydrogen production [1–5].

Indeed, the NG steam reforming ($CH_4 + H_2O = CO + 3H_2$) is an endothermic reaction ($\Delta H^0_{298K} = 206$ kJ/mol) and is limited by chemical equilibrium. This means that operation at high temperatures (850–900 °C) are required to reach significant hydrogen yields. In order to supply the reaction heat duty, a stream of NG is typically burned in the steam reforming furnace, thereby determining a reduction of the overall process efficiency as well as an increase in CO_2 emissions.

Under such a scenario, the coupling of the steam reforming unit with a Pd-based hydrogen-selective membrane can provide for the following benefits: (i) enhancement of hydrogen yield and process efficiency at low temperatures, since the continuous selective removal of hydrogen from reaction environment allows maintaining the gas mixture composition far from equilibrium one; (ii) replacement of the high-temperature flue gases used in the furnace with a cleaner energy source or with waste heat available from another process; (iii) use of cheaper steel alloys for the fabrication of the reforming tube instead of the expensive materials currently used to withstand the high operating temperatures of conventional steam reforming plants.

KT—Kinetics Technology has gained an impressive experience over the last 15 years in the design and operation of innovative pilot plants, membrane reformers-based, for the production of pure hydrogen [6–11]. In particular, a pilot unit of 20 Nm^3/h of pure hydrogen has been designed and tested for more than 3000 h in a relevant industrial environment with NG supplied from the town grid, with a basic configuration characterized by sequential steps of reaction and membrane separation, in order to achieve the overcoming of thermodynamic equilibrium, while maintaining at the same time slightly decoupled operating conditions for a reformer and a membrane module, thereby avoiding a great deal of thermal stress for Pd-based membranes. The former proposed configuration, even not totally in line with the process intensification concept, represents a more simplified membrane reactor concept, where the membrane reactor in principle does not suffer from engineering design challenges and accordingly might boost the industrial acceptance of the novel technology in the first phases of transition to it.

Over the last few years, with the aim to further reduce the CO_2 emissions, typical of the highly energy-intensive processes of the hydrocarbon industry, attention has been focused by the Company on the application of solar energy as a renewable energy source, with the possibility to exploit it with molten salts having the role of heating medium and energy vector. The integration between the solar plant and industrial processes is widely studied all over the world [12–15]. For instance, Concentrating Solar Power (CSP) offers an attractive option to power industrial-scale desalination plants requiring thermal and electrical energy. Integration of CSP with multi-effect distillation (MED) and Reverse Osmosis (RO) seems to be cheaper than photovoltaic (PV) systems [16]. Another potential application of solar energy is to power chemical processes when a high thermal duty is required: this is the case for endothermic reactions, which are a very promising opportunity for the production and storage of usable energy, and mitigation of CO_2 emissions.

Therefore, the aim of this paper is to report the experimental activities performed by the KT—Kinetics Technology-controlled company Processi Innovativi in the framework of the EU Project CoMETHy (Compact Multifuel Energy To Hydrogen conversion) in this field, studying the coupling of an integrated membrane reformer with solar-assisted molten salts heating, with the operation of a pilot unit specifically designed to this aim. The main feature of such a reactor is the complete integration of catalyst and membranes in only one vessel according to the process intensification concepts [17].

An economic analysis has been also carried out, in order to check the industrial feasibility of the option to provide for a competitive novel technology at reduced environmental impact.

2. Experimental: Pilot Plant Description

The pilot plant was designed for a capacity of 3 Nm^3/h of pure hydrogen. The facility is available at ENEA Casaccia premises, and the process scheme of the pilot unit is reported in Figure 1.

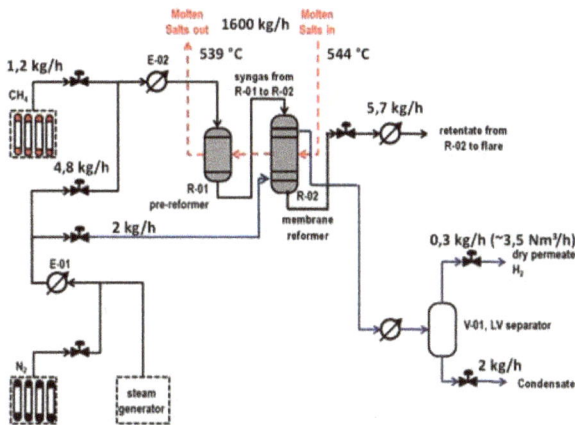

Figure 1. Solar energy-assisted membrane pilot plant process scheme.

The plant architecture is based on a first prereformer stage (R-01) placed upstream to an integrated membrane reactor (R-02). Methane is supplied by pressurized gas cylinders. A process steam and a sweep steam for membranes are produced with a dedicated electrical boiler. Reaction heat to R-01 and R-02 is supplied through a molten salts mixture fed to R-02 at a maximum inlet temperature of 550 °C and further routed to the R-01.

Molten salts are countercurrently flowing through the two reactors, entering first R-02 at the desired temperature and flow rate. Molten salts are stored in a tank and pumped at the desired flow rate through an electrical heater, where the desired temperature is applied before being sent to the chemical test section. Due to the relatively low thermal duty of the reactors, the molten salts return temperature from the test section was just about 10 °C lower than the molten salts supply temperature. For this reason, in order to maintain a temperature lower than 500 °C inside the molten salts tank, a fraction of the heat from the back stream was recovered by heating the molten salts stream from the tank, and finally the temperature was further reduced by means of an air cooler.

In the prereformer stage of R-01, a partial methane conversion occurs, and accordingly the produced syngas mixed with unconverted methane routed to R-02 contains a certain amount of hydrogen, allowing membrane in R-02 to be active just at the inlet of reactor.

R-01 was designed in a shell and tube configuration where molten salts mixture flows in the shell side supplying reaction heat, and catalysts are installed inside tubes (Figure 2).

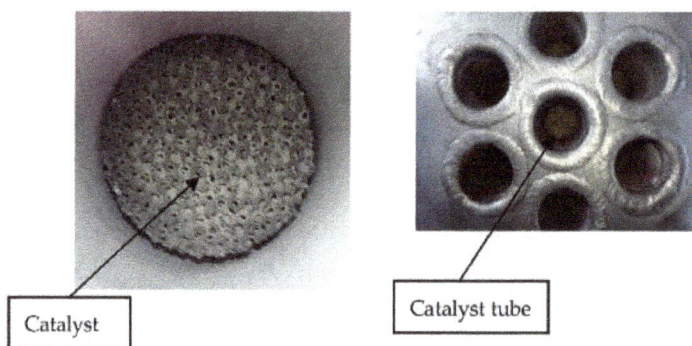

Catalyst

Catalyst tube

Figure 2. Catalyst arrangement in R-01 reactor.

R-02 was also designed in a shell and tube configuration (Figure 3).

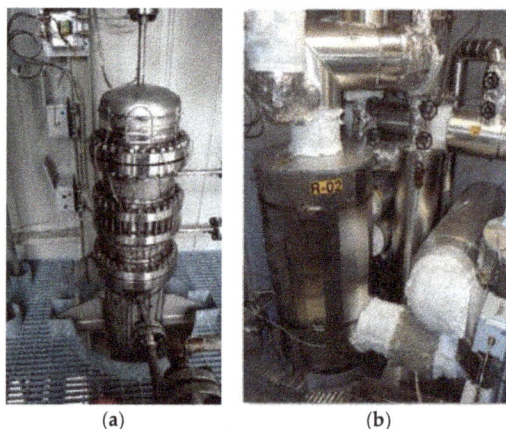

(a)

(b)

Figure 3. R-02 reactor. (a) overall reactor; (b) reactor installation.

The main difference with the R-01 reactor is that in particular in R-02, catalysts and membranes are arranged according to a tube-in-tube configuration with catalysts in the annular section around the membrane tube (Figure 4). The latter is equipped with an inner tube to allow for sweep gas flow on the permeate side.

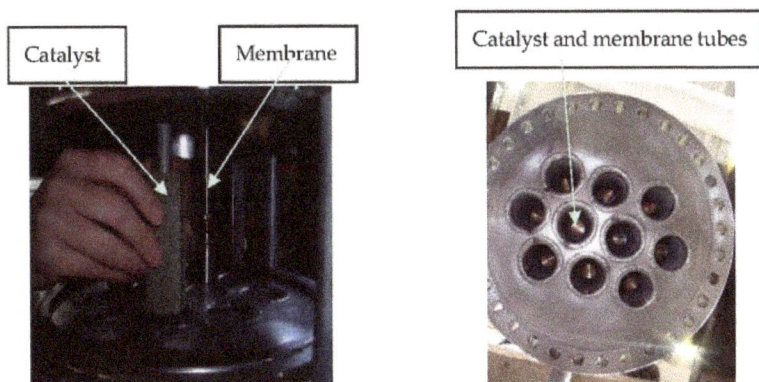

Figure 4. R-02 reactor membrane arrangement.

Ten Pd-based membrane tubes are installed in R-02, with a permeate stream collected from R-02 and cooled down in order to easily separate sweep gas as condensate.

The installed membranes were provided by ECN (Energy Research Centre of the Netherlands) and they had an outside diameter of 14 mm and a length of 80 cm for an overall area of about 0.35 m^2. The main characteristics of the membranes are reported in Table 1.

Table 1. Main characteristics of membranes installed in R-02.

Membrane	Supplier	Support	Membr. Selective Layer	Thick. Selective Layer, m	Membr. Area, m2	Geometry	Production Method	Permeance, Nm3/(hm2bar0.5) at 400–450 °C
M (R-02)	ECN	Al$_2$O$_3$	Pd	3–6	0.35	Tubular	Electroless Deposition	10–15

A bifunctional catalyst, nickel-noble metals-based, deposited on silicon carbide foam, is installed in both R-01 and R-02 reactors separately, shaped in the form of cylinder for R-01 (Figure 2) and annular cylinder for R-02 wrapped around the membrane tube (Figure 5).

Figure 5. R-02 reactor catalyst arrangement.

The catalysts were prepared at the ProCeed Lab of the University of Salerno. To improve membrane separation efficiency, a superheated steam is used as a sweep gas in a countercurrent configuration. Reactors R-01/R-02 and piping in contact with molten salts are electrical traced, in order to assure, during the start-up and shut down procedures, temperature above the salts freezing point.

The main operating conditions of the catalytic tests carried out with the integrated membrane reactor are reported in Table 2.

Table 2. Main operating conditions for the integrated membrane reactor.

Description	Value
Flow rate	
CH_4 inlet pre-reforming reactor R-01, kg/h	0.3–1.5
H_2O inlet pre-reforming reactor R-01, kg/h	4.5–6.0
H_2O Sweep gas, kg/h	0–2.0
Molten salts, kg/h	1250–1650
Pressure	
P feed side, barg	9.8–9.5
P permeate side, barg	0.4
Temperature	
T range molten salts, °C	450–550

For the calculation of heat and material balances of the pilot plant, each of these units was modeled with Aspen Plus 9.0 as a standard process simulator. For the integrated membrane reactor simulation, ASPEN Plus 9.0 was combined with the kinetic equations reported by Xu and Froment [18] and membrane permeation derived on Pd membranes coming from the same supplier and tested by the authors in the framework of previous experimental activities.

3. Results and Discussion

Different parameters were considered for the performance evaluation of the integrated system, such as the molten salts temperature, sweep steam flow rate and steam-to-carbon ratio. The effects of molten salts temperature on the methane conversion and the permeate H_2 flow rate at an operating pressure of 8.5 barg are reported in Figure 6.

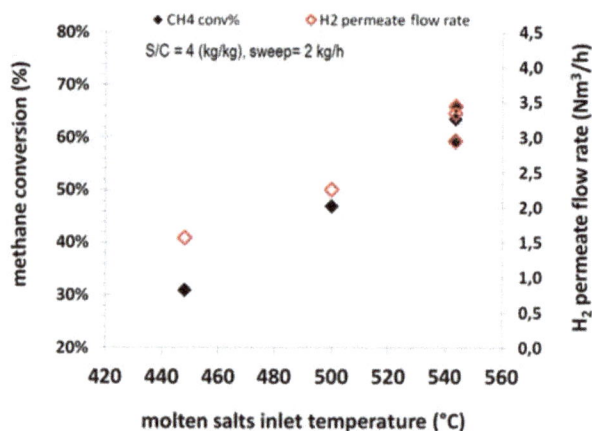

Figure 6. Effects of molten salts temperature on the methane conversion and the permeate H_2 flow rate.

It can be observed that by increasing the molten salts temperature, the methane conversion increases since the endothermic reforming reaction is promoted at high temperatures, with CH_4 conversion increasing from around 30% at 450 °C to around 60% at 540 °C. Obviously, the same trend can be observed for the flow rate of hydrogen permeating through the Pd membranes. At 540 °C, a permeate hydrogen flow rate of around 3.5 Nm^3/h of hydrogen can be observed. It is also of interest the low CO concentration can be detected on the permeate side. Such values in consequence of the low level in the retentate side are in order of 12 ppmv (dry basis) at 450 °C and about 50 ppmv (dry basis) at 540 °C. The increase of CO concentration on the permeate side is in line with the occurrence of water gas shift more promoted at lower temperatures, thereby reducing the CO content on the retentate side and accordingly the CO detected on the permeate side. In terms of permeate purity, all tests accounted for an average H_2 content higher than 99.8% mol (dry basis).

The effects of the sweep gas flow rate on the methane conversion and the permeate H_2 flow rate are reported in Figure 7 for two operating temperatures, 544 °C and 500 °C, respectively, at an operating pressure of 8.5 barg.

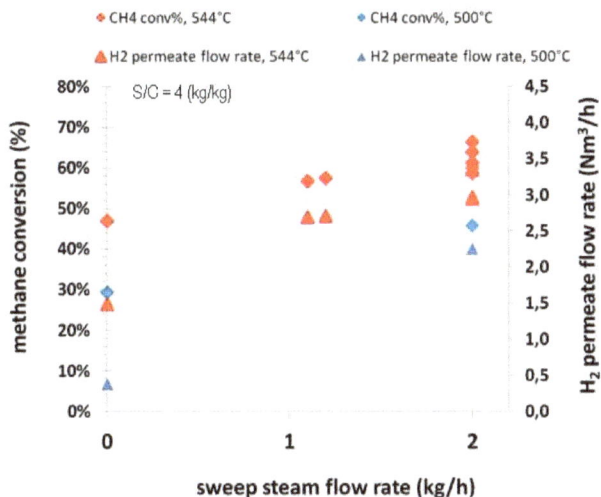

Figure 7. Effects of sweep gas flow rate on the methane conversion and the permeate H_2 flow rate.

For both operating temperatures, an increase in methane conversion and accordingly in the permeate H_2 flow rate can be observed with increasing of the sweep gas flow rate, since a higher value for the latter parameter enables a higher driving force across the membrane. However, it is important to observe that the strongest effect can be checked when the sweep gas flow rate is increased up to 1 kg/h. At higher flow rates, the impact is more negligible. At about 1 kg/h of sweep gas flow rate, the hydrogen recovery is around 70% at 544 °C. A productivity of H_2 of 1.5 Nm^3/h can be obtained, even in absence of sweep gas on the permeate side.

The effects of steam-to-carbon ratio on methane conversion and H_2 recovery are reported in Figure 8, at the three operating temperatures of 500 °C, 530 °C, 544 °C and an operating pressure of 8.5 barg. The sweep gas flow rate was kept constant at 2 kg/h.

An increase in methane conversion can be observed with increasing of the steam-to-carbon ratio in the feed. The effect is more pronounced at low temperatures. At the highest steam-to-carbon ratio investigated of 16 (on weight basis), a methane conversion of 99% can be detected at 544 °C. In terms of hydrogen recovery, in this condition, it is possible to achieve a recovery of more than 90%.

The performance comparison of the integrated membrane system with the thermodynamic equilibrium without a membrane is reported in Figure 9, at a feed pressure of 8.5 barg and a steam-to-carbon ratio of 4 (on weight basis).

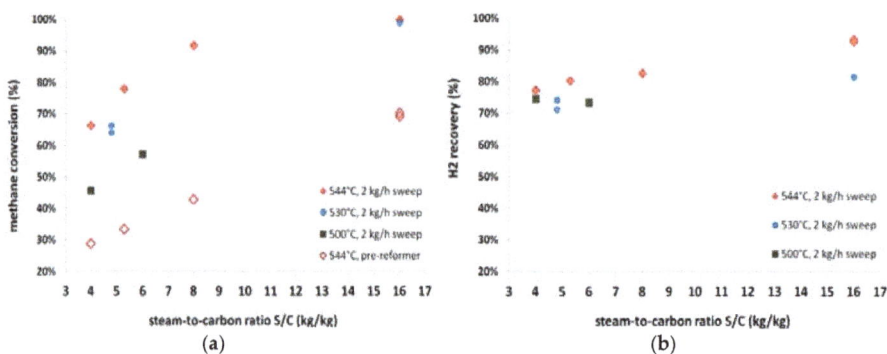

Figure 8. Effects of steam-to-carbon ratio on methane conversion (**a**) and H_2 recovery (**b**).

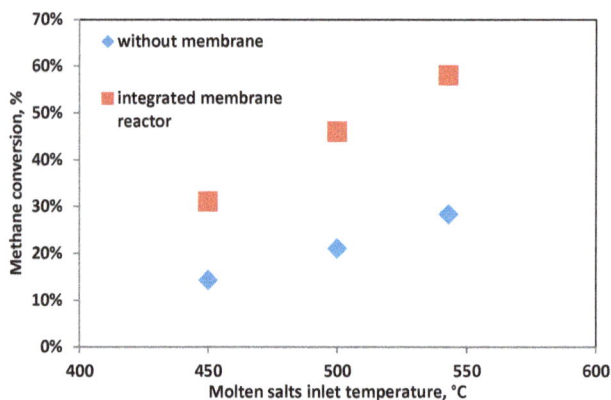

Figure 9. Methane conversion with an integrated membrane reactor and without a membrane.

An overall feed conversion of 58% can be achieved at 543 °C, doubling the conversion that can be achieved in a conventional reformer at the same temperature.

The system performance was also investigated in time-on-stream tests, in order to check its potential feasibility at industrial conditions. The results in terms of the methane conversion and the permeate hydrogen concentration are reported in Figure 10.

The system performance is very stable for more than 100 h of continuous operation and no macroscopical signs of reactor performance loss have been evidenced over the experimental operation period, despite handling of catalysts and membranes and the several switches of operative conditions. The achieved results confirm the potentiality of the solution for application at industrial conditions; even if this concept is further assessed, it would be important to check the behavior of the membrane-catalyst coupling for at least 1000 h of continuous operation. Indeed, this order of magnitude of stability time is usually required for the catalyst at an accepted industrial level. In this way, if the stability of the membrane and that of catalyst are aligned to such order of magnitude, this allows to avoid too many frequent shut down operations of the plant for any procedure of maintenance or replacement of key components.

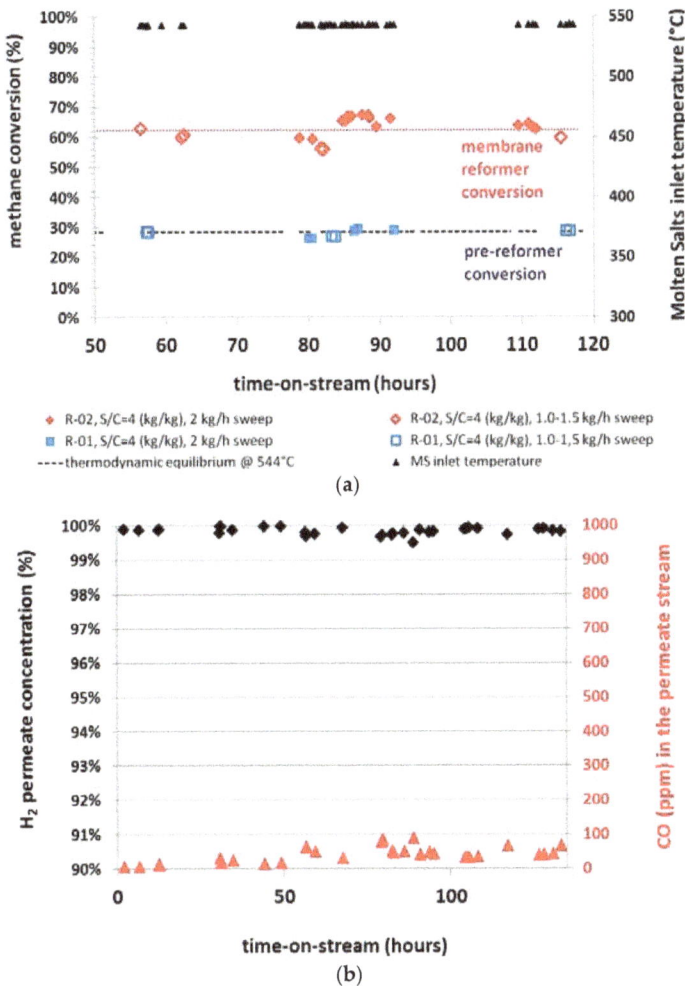

Figure 10. Methane conversion (a) and permeate H_2 concentration (b) in time-on-stream tests.

4. Economic Analysis

In order to make a preliminary economic analysis about the potentiality of the solar energy coupling with membrane reactors at an industrial capacity of 5000 Nm^3/h of hydrogen, for the sake of simplicity, it was assumed to operate with a six-step membrane reactor in an open architecture, where reactions stages are followed by membrane stages.

A process arrangement has been studied in order to realize a high-energy-efficiency process coupled with a low hydrogen production cost and CO_2 removal, based on the work proposed by Atsonios et al. [19]. The process scheme considered with CO_2 capture in pre-combustion is reported in Figure 11. The main units are represented by: (i) the multistep membrane steam reformer, (ii) product compression, (iii) power island. Each of these units was modeled with a standard process simulator.

The rate of hydrogen removed with a Pd/Ag membrane is described by the Hydrogen Recovery Factor (HRF), which is strongly correlated with the total installed membrane area, the pressure at the permeate side, and that at the retentate side. The pressure at the feed side was set to be equal to 9.2 barg, and meanwhile, the permeate side was kept at 0.7 barg. A low-pressure steam extracted from the steam cycle was used as a sweep gas to reduce hydrogen partial pressure. A sweeping steam was added in order to have an equimolar mixture of hydrogen and steam on the permeate side. The HRF (Hydrogen Recovery Factor) was kept constant in our simulation and equal to 90% (overall).

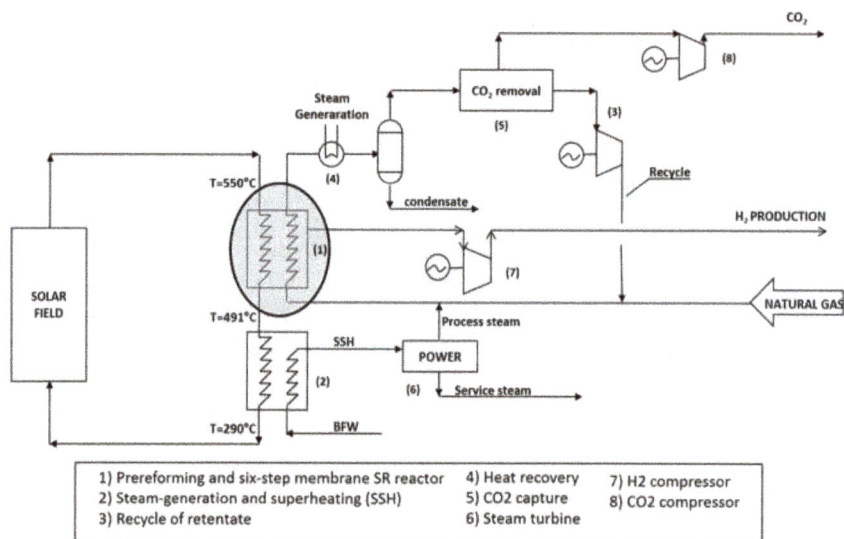

Figure 11. Process scheme for membrane steam reforming solar energy assisted with export power production.

The solar field was designed for heat molten salts with a temperature up to 550 °C. Molten salts are fed to a steam reforming section composed by a pre-reformer and a six-step membrane Steam Reforming reactor. The outlet temperature of molten salts exiting the reaction section is 490 °C, so they can be used to produce the process steam required for the steam reforming reaction and the sweeping steam to be used at the membrane separation stage: additional steam is produced and used to generate power in a steam turbine.

Excess power is then exported. Molten salts are then forwarded to the solar field at a temperature of 290 °C. The NG is mixed with steam and sent to the reaction section; the produced hydrogen is recovered by membranes and then compressed to a pressure of 20 barg. The retentate is cooled down to generate steam (heat recovery) and then, after a CO_2 removal stage through a conventional amine unit, compressed and recirculated to the SR section. In this way, such a scheme achieves the complete conversion of NG in H_2 and CO_2. The total duty provided by the solar field is 16.7 MMkcal/h, much larger than what is required by the reforming section. The specific export of power was calculated in 0.4 kWh/Nm³ of produced hydrogen for an overall production of 2 MWh. The membrane area was estimated at 976 m². In the calculation, a 5 μm thick larger of Pd/Ag23 wt% with a permeability of 30 Nm³/h m² bar$^{0.5}$ and H_2/CO minimum selectivity of 200 was assumed [6].

The cost of production (COP) per Nm³ of H_2 produced was calculated by adding the capital expenditure (CAPEX) and the operating expenditure (OPEX) costs. No benefit was taken for CO_2 reduction emission. In addition, it is assumed that SR reactor is powered with a molten salts flow heated with solar energy for 5000 h/year, and in the remaining period of time, 3400 h/year, molten salts are heated through a process heater where NG is fired. Parameters used for economic analysis are reported in Table 3.

Table 4 reports, together with operating conditions, the relevant COP for the innovative scheme compared with a conventional steam reformer scheme, where reaction duty is provided by purge gas from pressure swing adsorption (PSA) and additional fuel gas. The plant architecture is that it provides the steam required by the reboiler of the CO_2 recovery unit.

Table 3. Key economic parameters for the economic evaluation.

Description	Values
NG cost, €/kg	0.22
Electricity cost, €/kWh	0.075
Annual depreciation factor	10
N° of hours powered by sun	5000

Table 4. Cost of production (COP) of H_2 for the conventional scheme and molten salts-based steam reforming.

Description	Conventional SR	Molten Salts-Based SR	
H_2 production, Nm^3/h	5000	5000	5000
N of hours	8400	5000	3400
Feed + Fuel, ton/h	1.6	0.98	2.57
Reformer duty, MMkcal/h	5.0	3.9	3.9
Solar field duty, MMkcal/h	-	16.7	-
H_2 compression at 20 barg, kWh/Nm^3 H_2	-	Included	Included
CO_2 recovery, ton/h	2.55	2.55	2.55
OPEX, €/Nm^3 H_2 (calculated)	0.077	0.045	0.115
CAPEX, M€	12.5	22.5	22.5
Export of power, kWh/Nm^3 H_2	-	0.4	0.4
Depreciation rate per year, M€	1.25	2.3	2.3
Depreciation rate, M€/Nm^3 H_2	0.030	0.054	0.054
Total production costs, €/$Nm^3 H_2$	0.106	0.068	0.138
Total average production cost, €/Nm^3 H_2	0.106	0.097	
Total production cost comparison	100	91	

It is quite evident that the innovative scheme is interesting and that the membrane assisted H_2 production powered by solar energy, and coupled with CO_2 removal may already compete with conventional steam reforming if electricity price is high enough.

In this regard, by increasing the electricity price, the difference in COP between the conventional scheme and the molten salts-based SR is higher at a higher electricity price. Anyway, as shown in Figure 12, the two configurations become equivalent at an electricity price much lower than current values.

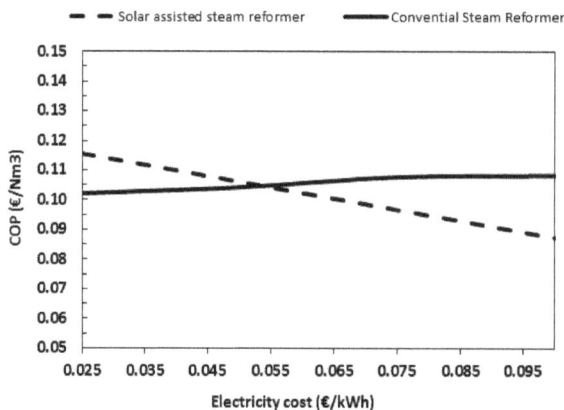

Figure 12. Influence of COP vs. electricity cost.

It is also important to note that in conventional technology, about 70% of the production costs are related to the variable costs. The situation is quite different for the innovative scheme, where more than 55% of COP is related to the CAPEX and then only 45% is related to OPEX. Such a difference implies that, if accelerated depreciation is applied, production costs of a scheme will be more affected than for the conventional SR technology. In particular, it is calculated that the break-even point is reached at a depreciation factor of 7; with a depreciation factor of 12, the difference becomes even more important, reaching almost 15%.

5. Conclusions

The performance of a compact membrane reactor for steam reforming coupled with solar energy has been evaluated at a pilot level. The membrane reactor has been designed to integrate structured catalysts and Pd-based membranes with an overall thermal sustainment provided by solar heated molten salts. The design activities show that the engineering of a membrane reactor involves the proper optimization and arrangement of catalyst volume and membrane area. The experimentation carried out clearly indicates that the developed system is able to guarantee a high-purity hydrogen stream, keeping low feed consumption. In addition, a preliminary economic analysis, aiming at evaluating the benefits from the integration between solar energy and membrane steam reforming in open architecture, shows that the solution is a promising approach to minimizing the energy penalty, the hydrogen production cost and CO_2 emitted per ton of product. Experimental time-on-stream tests with longer duration than the reported ones in this paper would further be helpful to definitely consider this scheme potentially applicable at an industrial level.

It is also worth mentioning that a proper design of a thermal storage as well as that of a back-up system could allow managing properly fluctuation of solar energy in order to assure continuous operation for the hydrogen plant.

6. Patents

Iaquaniello, G., Salladini, A., Morico, B. Method and system for the production of hydrogen. US Patent US9493350B2, 15 November 2016 (priority date 16 March 2012).

Author Contributions: All authors contributed to writing and correcting the paper.

Funding: The research leading to these results has received funding from the European Union's Seventh Framework Programme (FP7/2007-2013) for the Fuel Cells and Hydrogen Joint Technology Initiative under grant agreement number 279075 for CoMETHy project.

Conflicts of Interest: The authors declare no conflicts of interest.

References

1. Alique, D.; Martinez-Diaz, D.; Sanz, R.; Calles, J.A. Review of Supported Pd-Based Membranes Preparation by Electroless Plating for Ultra-Pure Hydrogen Production. *Membranes* **2018**, *8*, 5. [CrossRef] [PubMed]
2. Van Delft, Y.C.; Overbeek, J.P.; Saric, M.; de Groot, A.; Dijkstra, J.W.; Jansen, D. Towards application of palladium membrane reactors in large scale production of hydrogen. In Proceedings of the 8th World Congress on Chemical Engineering, Montreal, QC, Canada, 23–27 August 2009.
3. Basile, A. Hydrogen Production Using Pd-based Membrane Reactors for Fuel Cells. *Top. Catal.* **2008**, *51*, 107. [CrossRef]
4. Fernandez, E.; Helmi, A.; Medrano, J.A.; Coenen, K.; Arratibel, A.; Melendez, J.; de Nooijer, N.C.A.; Spallina, V.; Viviente, J.L.; Zuñiga, J.; et al. Palladium based membranes and membrane reactors for hydrogen production and purification: An overview of research activities at Tecnalia and TU/e. *Int. J. Hydrog. Energy* **2017**, *42*, 13763–13776. [CrossRef]
5. Shu, J.; Grandjean, B.P.A.; Van Neste, A.; Kaliaguine, S. Catalytic palladium-based membrane reactors: A review. *Can. J. Chem. Eng.* **1991**, *69*, 1036–1060. [CrossRef]
6. De Falco, M.; Iaquaniello, G.; Salladini, A. Experimental tests on steam reforming of natural gas in a reformer and membrane modules (RMM) plant. *J. Membr. Sci.* **2011**, *368*, 264–274. [CrossRef]

7. De Falco, M.; Salladini, A.; Iaquaniello, G. Reformer and membrane modules (RMM) for methane conversion: Experimental assessment and perspectives of said innovative architecture. *ChemSusChem* **2011**, *4*, 1157–1165. [CrossRef] [PubMed]

8. Iaquaniello, G.; Palo, E.; Salladini, A.; Cucchiella, B. Using palladium membrane reformers for hydrogen production. In *Palladium Membrane Technology for Hydrogen Production, Carbon Capture and Other Applications*; Doukelis, A., Panopoulos, K., Koumanakos, A., Kakara, E., Eds.; Woodhead Publishing: Sawston, UK, 2014; pp. 287–301. ISBN 9781782422341.

9. Falco, D.M.; Iaquaniello, G.; Salladini, A. Steam reforming of natural gas in a reformer and membrane modules test plant: Plant design criteria and operating experience. In *Membrane Reactors for Hydrogen Production Processes*; De Falco, M., Marrelli, L., Iaquaniello, G., Eds.; Springer: Berlin, Germany, 2011; pp. 201–224. ISBN 978-0-85729-150-9.

10. Iaquaniello, G.; Palo, E.; Salladini, A. Application of membrane reactors: An industrial perspective. In Proceedings of the MR4PI2017, Villafrance de Verona, Italy, 9–10 March 2017.

11. Salladini, A.; Iaquaniello, G.; Palo, E. Membrane Reforming Pilot Testing: KT Experiences. In *Membrane Reforming Pilot Testing: Applications for a Greener Process Industry*; Wiley: Hoboken, NJ, USA, 2016; ISBN 978-1-118-90680-4.

12. Valenzuela, C.; Mata-Torres, C.; Cardemil, J.M.; Escobar, R.A. CSP + PV hybrid solar plants for power and water cogeneration in northern Chile. *Sol. Energy* **2017**, *157*, 713–726. [CrossRef]

13. Leiva-Illanes, R.; Escobar, R.; Cardemil, J.M.; Alarcón-Padilla, D.C.; Uche, J.; Martínez, A. Exergy cost assessment of CSP driven multi-generation schemes: Integrating seawater desalination, refrigeration, and process heat plants. *Energy Convers. Manag.* **2019**, *179*, 249–269. [CrossRef]

14. Islam, M.T.; Huda, N.; Abdullah, A.B.; Saidur, R. A comprehensive review of state-of-the-art concentrating solar power (CSP) technologies: Current status and research trends. *Renew. Sustain. Energy Rev.* **2018**, *91*, 987–1018. [CrossRef]

15. Boudries, R. Techno-economic study of hydrogen production using CSP technology. *Int. J. Hydrog. Energy* **2018**, *43*, 3406–3417. [CrossRef]

16. Iaquaniello, G.; Salladini, A.; Mari, A.; Mabrouk, A.A.; Fath, H.E.S. Concentrating solar power (CSP) system integrated with MED-RO hybrid desalination. *Desalination* **2014**, *336*, 121–128. [CrossRef]

17. Morico, B.; Gentile, A.; Iaquaniello, G. Molten Salt Solar Steam Reforming: Process Schemes Analysis. In *Membrane Reactor Engineering Applications for a Greener Process Industry*; Basile, A., De Falco, M., Centi, G., Iaquaniello, G., Eds.; Wiley: Hoboken, NJ, USA, 2016; ISBN 9781118906804.

18. Xu, J.; Froment, G.F. Methane steam reforming, methanation and water gas shift: I. Intrinsic kinetics. *AIChE J.* **1989**, *35*, 88–96. [CrossRef]

19. Atsonios, K.A.; Panopoulos, K.D.; Doukelis, A.F.; Koumanakos, A.K.; Morud, J.; Kakaras, E. Natural gas upgrading through hydrogen selective membranes: Application in carbon free combined cycles. *Energy Procedia* **2013**, *37*, 914–923. [CrossRef]

chemengineering

MDPI

Article

Dry Reforming of Methane in a Pd-Ag Membrane Reactor: Thermodynamic and Experimental Analysis

Alessio Caravella [1,2]ⓘ, **Adele Brunetti** [1,*]ⓘ, **Monia Grandinetti** [1,2] **and Giuseppe Barbieri** [1]

[1]   Institute on Membrane Technology, National Research Council (ITM-CNR), Via P. Bucci, Cubo 17C, 87036 Rende (CS), Italy; alessio.caravella@unical.it (A.C.); monia.grandinetti@hotmail.it (M.G.); g.barbieri@itm.cnr.it (G.B.)

[2]   Department of Environmental and Chemical Engineering (DIATIC), University of Calabria, Via P. Bucci, Cubo 44A, 87036 Rende (CS), Italy

*   Correspondence: a.brunetti@itm.cnr.it; Tel.: +39-0984-492012; Fax: +39-0984-402103

Received: 13 September 2018; Accepted: 9 October 2018; Published: 10 October 2018

Abstract: The present work is a study of CO_2 Reforming of Methane (DRM) carried out in a catalytic Pd-based membrane reactor. A detailed thermodynamic analysis is carried out, calculating the chemical equilibrium parameters in two different cases: (a) DRM along with the Reverse Water Gas Shift (RWGS) reaction and (b) DRM along with both RWGS and the Boudouard Reaction (BR). The performance of membrane reactor is then experimentally analyzed in terms of methane conversion, hydrogen recovery and H_2/CO reaction selectivity by varying feed pressure and CO_2/CH_4 feed molar ratio and 500 °C and GHSV = 100 h^{-1}. Among the obtained results, a CH_4 conversion of about 26% and a H_2 recovery of 47% are achieved at low feed pressures, exceeding the traditional reactor equilibrium conversion. This effect can be attributed to the favorable thermodynamics coupled to the hydrogen permeation through the membrane. This study further demonstrates the general effectiveness of membrane-integrated reaction processes, which makes the production of syngas more efficient and performing, providing important environmental benefits.

Keywords: membrane engineering; hydrogen production; CO_2 conversion

1. Introduction

In the last decade, the energy demand has been growing by 1.2% a year and fossil fuels still maintain a production share of ca. 75%. However, the ever-stricter problems connected to a sustainable growth and to a lower environmental impact lead to the conclusion that the time of easy oil consumption is finished. Nowadays, the necessity to release energy production from oil and natural gas as primary energy sources is becoming more and more pressing [1–3]. Indeed, more in general, diversifying such sources in order to assure supply, and in the meantime increase effort dedicated to the reduction of environmental problems, has led to the development of alternative technologies designed to enhance both the efficiency and environmental acceptability of energy production, storage and use, in particular for power generation [4]. Among these technologies, the exploitation of light hydrocarbons is surely the main realistic energy source, since they allow both power generation and environmentally-friendly fuel production [5,6].

Actually, converting CO_2 into valuable hydrocarbons seems to be one of the most recent advances in CCU (Carbon Capture and Utilization), being one of the best solutions to both global warming and energy lacking problems. Several technologies have recently been explored and are reported in literature for CO_2 conversion [7]. These technologies are based on hydrogenation, electrochemical, thermochemical or biocatalytic processes, and photocatalytic reduction. Among these, photocatalytic conversion is growing faster in the development not only of more active catalysts but also in the design of innovative process units [8,9]. In addition, biochemical and bio-mimetic approaches are also reaching interesting results although they are still to be applied at large scale. [10,11]. A very active research area is the development of an "artificial leaf" [12] that collects energy in a similar way as a natural one [13,14], combining water oxidation and CO reduction to produce liquid fuels by artificial photosynthesis; however, the development of this technology is also far from real scale, owing to limitations on solar energy-to-chemical conversion efficiency, costs, robustness and of easy construction [13].

Dry reforming of methane (DRM) is a reaction that has led significant interest owing to the possibility to convert greenhouse gases, CO_2 and CH_4 to produce valuable products. The resulting synthesis gas has, in fact, a CO/H_2 ratio close to 2, which is more appropriate for forming hydrocarbons by Fischer-Tropsch synthesis and for carbonylation and hydroformylation reactions [15]. Moreover, natural gas and biogas from fields having high carbon dioxide content can be directly used for reaction, avoiding separation and purification stages. Although the undoubted benefits, DRM development on large scale is still limited owing to the usual rapid deactivation of catalysts due to coke formation and the occurrence of side reactions, which decrease the yield of syngas [16–19].

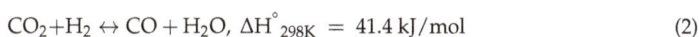

$$CH_4 + CO_2 \leftrightarrow 2CO + 2H_2, \ \Delta H^{\circ}_{298K} \ = \ 247 \ kJ/mol \tag{1}$$

$$CO_2 + H_2 \leftrightarrow CO + H_2O, \ \Delta H^{\circ}_{298K} \ = \ 41.4 \ kJ/mol \tag{2}$$

$$2CO \leftrightarrow CO_2 + C, \ \Delta H^{\circ}_{298K} \ = \ -172.4 \ kJ/mol \tag{3}$$

A common overbearing problem is the reverse water-gas shift (RWGS) associated with dry reforming, which consumes the hydrogen produced by the reaction to form water. This is much more evident at high pressure, because the reaction (2) is favoured with respect to reaction (1) owing to the much higher reactivity of H_2 over CH_4 [20].

Membrane reactors (MRs) are a promising solution to overcome these limitations, combining the reaction and H_2 separation by means of a selective membrane. The presence of a hydrogen-selective membrane allows the removal of hydrogen from reaction side with a contemporary recovery a hydrogen rich/pure stream [21] and the shifting of the reaction toward further conversion. In addition, as the permeation is a pressure driven process, the negative effect of reaction pressure, which favors RWGS, is counterbalanced by the promotion of hydrogen removal from reaction volume. In addition, MR operates below 600 °C, thus, below the temperature range at which coke deposition readily occurs [22]. In most cases, the MRs used for DRM are constituted by a selective membrane—usually Pd-based—having only the separating function, whereas the catalyst is packed in the annulus between the membrane and reactor shell [23–29].

As it is well known, Ni-based is the most used commercial catalyst for DRM. However, it suffers from severe loss of catalytic activity mainly due to the coke formation. Industrial steam reformers are fed with steam-to-carbon ratios close to 3 to suppress the coking, resulting in less efficient operation. To enable operation with lower steam-to-carbon ratios, new catalysts need to be developed that are simultaneously highly active, resistant to coking, and low cost. Such catalysts can be based on novel materials based on platinum group metals (Ru, Rh, Pt, Pd, etc.) [30,31]. Very recently, Simak and Leshkov [32] demonstrated the advantages obtained by using a 0.15 wt % Ru/-Al_2O_3 catalyst in methane steam reforming coupled with a Pd-Ag membrane (5 micrometer thick). This MR, operated at 650 °C and 8 bar of feed pressure, showed high catalytic activity with a methane conversion higher than 50% and hydrogen yield of about 70%. At the same time, the MR showed stable performance over a total of 400 h on stream, including operation with low steam-to-carbon ratios of 1 and 2, and combined dry-steam reforming. On the basis of these positive results, we decided to carry out DRM in a Pd-Ag MR at 500 °C in the presence of a 0.5 wt % Ru/Al_2O_3 catalyst. As also observed by Simak and Leshkov [32], conversions achievable are quite low (<50%) owing to RWGS and the coke formation. This latter can affect not only the catalyst activity but also the membrane stability. For this purpose, we decided to use a Pd-Ag commercial membrane (100 micron thick), which exhibits high chemical and mechanical resistance, good permeability and infinite H_2 selectivity [21], circumventing the various issues that usually could affect ultrathin membranes. The coupling of the selective hydrogen removal offered by the membrane with the stability of an Ru-based catalyst could provide interesting insight in DRM reaction development.

CH$_4$ conversion, hydrogen recovery and reaction selectivity were analyzed as a function of feed pressure and CO$_2$/CH$_4$ feed molar ratio. In addition, the experimental analysis was coupled with a detailed thermodynamic study of DRM reaction, calculating the chemical equilibrium parameters in two different cases: (a) DRM along with the Reverse Water Gas Shift (RWGS) reaction, and (b) DRM along with both RWGS and the Boudouard Reaction.

2. Materials and Methods

The experiments were carried out in a tube in tube MR (Figure 1) where the outer tube is a stainless-steel shell, whereas the inner tube is the Pd-alloy self-supported membrane, blind on one end. The catalytic bed was 0.5% Ru/Al$_2$O$_3$ commercial catalytic pellets, packed in the annulus and the permeated stream is recovered in the core of the membrane (permeate side). The sealing between the membrane and the reactor shell was obtained with a graphite O-ring via compression. Both the membrane and MR characteristics are summarized in Table 1.

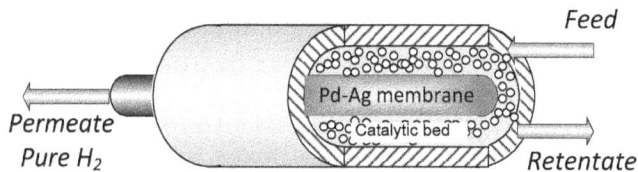

Figure 1. Membrane reactor scheme.

Table 1. Membrane reactor characteristics.

Membrane	Pd-Ag Commercial (Goodfellow) Self-Supported
Thickness	100 micrometers
Superficial Area	3 cm^2
Outer Diameter	1 mm
Length	93 mm
Catalyst Weight	8.5 g

The laboratory plant assembled to perform the present investigation is sketched in Figure 2.

The reactor was placed inside an electric furnace to keep the desired temperature. The gas mixture was fed into the module by two mass flow controllers (Brooks Instrument 5850S, Hatfield, PA, USA). The outlet streams were fed to two bubble soap flow meters in order to measure the gas flow rates and, thus, to evaluate the permeating flux. A pressure gauge with a backpressure controller was placed on the retentate stream to keep and measure the feed pressure; whereas permeate pressure was regulated by a vacuum pump. The retentate and permeate streams compositions were analyzed by means of a gas chromatograph (Agilent 7890N, Santa Clara, CA, USA) with two parallel analytical lines. Each line was equipped with two columns: An HP-Plot-5A (to separate permanent gases such as H$_2$, N$_2$ and CO) and an HP-Poraplot-Q (for other species) allowing the retentate and permeate streams to be analyzed at the same time. The temperature was measured by using a thermocouple positioned in the middle of the reactor shell (inside the catalyst bed). After assembly, the membrane reactor was charged under nitrogen pressure at 8 bar on the feed side, closing the retentate. No pressure falls were detected after one hour, confirming the absence of leakages. The same procedure was repeated at 500 °C, before starting reaction measurements. Before reaction experiments, the membrane was assembled in the module without packing the catalyst and permeation measurements were carried out with H$_2$ single gases at different temperatures and pressures. For this purpose, the feed-side absolute pressure was set at 4, 6 and 8 bar, whilst the permeation-side pressure was fixed at atmospheric pressure during each series of permeation measurements. Afterwards, a CO_2:CH_4 stream was fed for reaction experiments at 500 °C, analyzing, in particular, the effects of the feed molar ratio, and the feed and permeate pressures. A summary of the operating conditions are reported in Table 2.

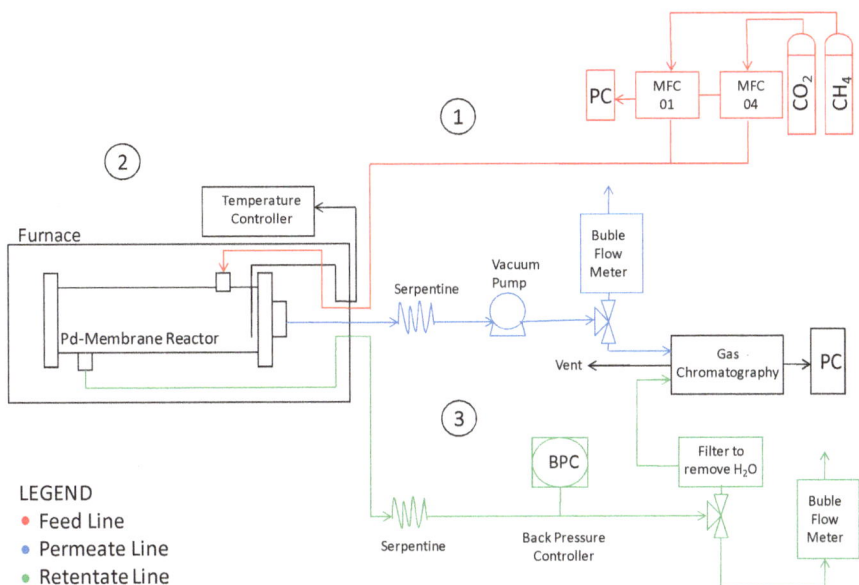

Figure 2. Sketch of the lab-scale plant for permeation and reaction experiments.

Table 2. Operating conditions for experimental measurements.

Temperature, °C	Permeation	400, 450, 500
	Reaction	500
Pressure, bar	Feed	1–8
	Permeate	0.02; 1
CO_2/CH_4 Feed Molar Ratio		1, 1.5
GHSV, h^{-1}		100

Generally, as also done in the present work, the H_2 permeating flux through Pd-alloy membranes can be described by Sieverts' law (Equation (4)) when the diffusion through the metal bulk is the rate-determining step. Under these conditions, the hydrogen permeating flux is considered a linear function of the permeation driving force, which is given by the difference of the square root of the H_2 partial pressure on both membrane sides. This assumption has been done since, as it can be seen in the next section, the permeating flux is fully linear with the square root of the H_2 partial pressure on both membrane sides.

$$H_2 \ permeating \ flux = Permeance^0_{H_2} \ e^{-E/RT} \left(\sqrt{P_{H_2}^{Reaction \ Side}} - \sqrt{P_{H_2}^{Permeate \ Side}} \right), \ mol{\cdot}m^{-2}{\cdot}s^{-1} \quad (4)$$

As for the reactor performance, CH_4 conversion in both TR and MR was calculated using Equation (5) including CH_4 present in the feed and retentate streams. In particular, the conversion was calculated as the average value between the CH_4 fed to the MR and that detected in the retentate. Each value of conversion reports an error bar taking into account the carbon balance. It was calculated for each measurement and it is comprised in the range -8.7% to 5%.

$$CH_4 \; conversion = \frac{CH_4^{Feed} - CH_4^{Retentate}}{CH_4^{Feed}}, \; - \tag{5}$$

The recovery capability of the MR was quantified in terms of H_2 recovery (Equation (6)), which is defined as the H_2 fraction permeated through the membrane with respect to all of the H_2 present in both outlets of MR.

$$H_2 \, Recovery \; = \; \frac{F_{H_2}^{Permeate}}{F_{H_2}^{Permeate} + F_{H_2}^{Retentate}}, \; - \tag{6}$$

Catalyst is periodically regenerated with a diluted stream of 10% H_2 in argon after each set of reaction experiments. The reverse methane decomposition is an exothermic reaction favoured at low temperatures. If coke is present, it would react with hydrogen to produce methane. Based on this fact, it is possible to calculate the coke reacted from the methane produced. For this purpose, the retentate stream was analyzed by gas chromatography, measuring at the same time the outlet flow rate by bubble flow meter. From such methanation tests, the total coke content in the catalyst after the reaction experiment was calculated from the amount of CH_4 formed and crosschecked with carbon balance calculations. Each measurement was repeated three times, alternated by methanation, for a total experimental campaign that lasted about 500 h. The measurements are reproducible, confirming a good stability of catalyst for the whole period of experiments.

3. Results and Discussion

3.1. Thermodynamic Analysis

The understanding of the thermodynamic behavior of DRM for syngas and hydrogen production is important to determine the most favorable reaction conditions. The theoretical background of thermodynamic analysis is reported in Appendix A. The thermodynamic equilibrium of DRM along with some side-reactions is evaluated by minimization of the total Gibbs free energy, which is carried out in the MATLAB® environment (see Appendix A for calculation details). As mentioned above, the independent reactions considered are the following: The DRM, RWGS and Boudouard reaction. More specifically, the analysis of the Boudouard reaction is important because coke is formed by it. To the best of our knowledge, there are no references in the open literature about equilibrium calculation in the co-presence of coke.

3.1.1. Calculation Validation

To verify the accuracy of the calculated values, the equilibrium of DRM and reverse WGS reaction was preliminarily studied comparing the obtained results with some present in the literature in terms of CH_4 conversion and equilibrium constant. As shown in Figure 3 (equilibrium conversion) and in Table 3 (equilibrium constant), the results from both MATLAB code and literature data agree very well.

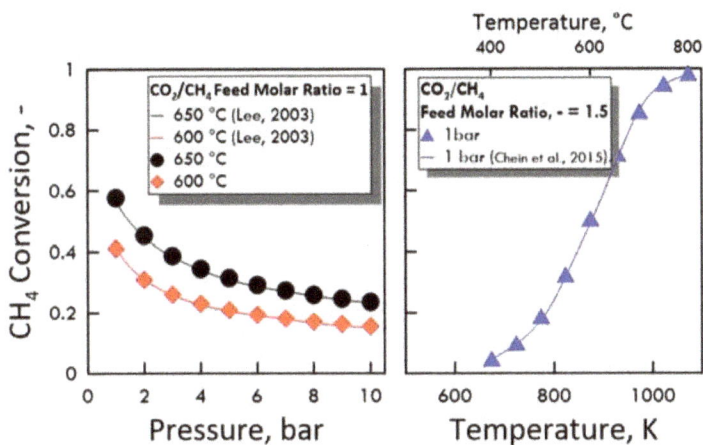

Figure 3. CH_4 conversion at equilibrium: (**Left** side) as a function of pressure, CO_2/CH_4 feed molar ratio of 1, temperature of 650 °C (●) and 600 °C (◆), compared with experimental data (Lee (2003) [33]); (**right** side) as a function of temperature, CO_2/CH_4 feed molar ratio of 1.5, pressure of 1 bar (▲), compared with experimental data (Chein et al. (2015) [34]).

Table 3. Equilibrium constant for dry reforming of methane (DRM) and reverse water gas shift (RWGS) reactions calculated at a temperature of 600 °C and 650 °C and compared with experimental data of Lee (2003) [33].

	Equilibrium Constant K_{eq}			
	DRM		RWGS	
	Lee (2003) [33]	MATLAB code	Lee (2003) [33]	MATLAB code
600 °C	0.19	0.20	0.37	0.40
650 °C	1.31	1.38	0.48	0.53

3.1.2. Equilibrium Calculation: DRM and RWGS

After ensuring the MATLAB method correctness, both CO_2/CH_4 feed molar ratio and pressure effect were evaluated on the methane equilibrium conversion. Figure 4 shows the CH_4 equilibrium conversion as a function of temperature and varying CO_2/CH_4 feed molar ratio for 1 bar (left side) and 10 bar (right side). At a pressure value sets, CH_4 conversion increases with CO_2/CH_4 feed molar ratio and with temperature. The DRM reaction is endothermic, and is favored at high temperature. Figure 5 shows the CH_4 equilibrium conversion as a function of temperature and varying pressure for CO_2/CH_4 feed molar ratio of 1 (left side) and 2 (right side). As pressure increases, CH_4 conversion decreases. The DRM reaction occurs with the increase in molar number and is unfavorable at high pressures. The pressure negative effect is due to the reaction thermodynamics, according to the Le Chatelier-Braun principle. In particular, methane conversion approaches limiting values as the pressure increases.

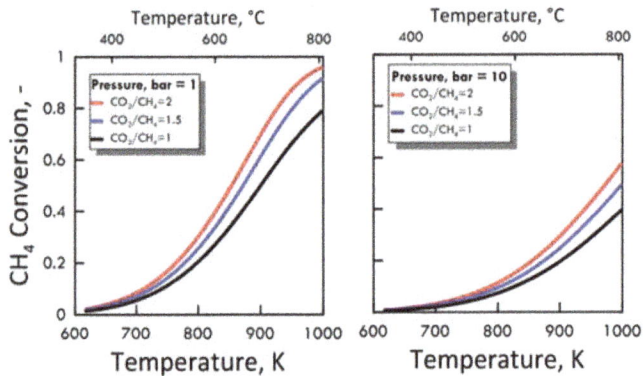

Figure 4. CH$_4$ equilibrium conversion at 1 bar (**left** side) and 10 bar (**right** side) as a function of temperature, varying CO$_2$/CH$_4$ feed molar ratio from 1 to 2.

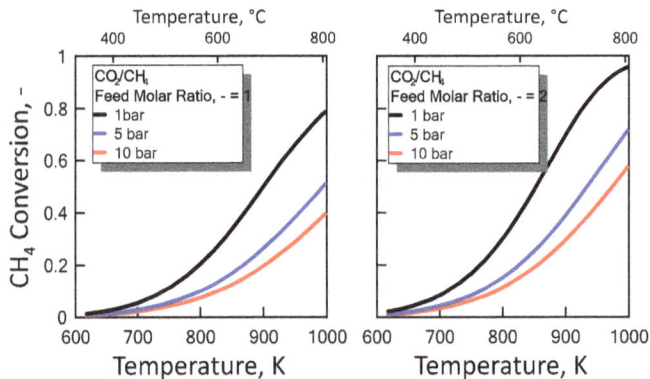

Figure 5. CH$_4$ equilibrium conversion at CO$_2$/CH$_4$ feed molar ratio of 1 (**left** side) and 2 (**right** side) as a function of temperature, varying the pressure from 1 to 10 bar.

3.1.3. Equilibrium Calculation: DRM, Reverse WGS and Boudouard Reactions

Carbon formation is one of the basic phenomena when hydrocarbon species are involved in chemical reactions. It is therefore interesting to examine its effect on the reaction. Figure 6 shows the equilibrium constants variation of the reactions involved as a function of temperature. For a strong endothermic reaction, DRM equilibrium constant increases dramatically with increasing reaction temperature. Thus, high conversion is favored at high temperatures. The equilibrium constants of the moderate endothermic reactions, the reverse WGS reaction, also increase with temperature. The carbon deposition by Boudouard reaction is exothermic and thermodynamically unfavorable at high temperatures. Therefore, high reaction temperatures are more favorable to increasing the equilibrium conversion of the DRM reaction than that of the side reactions. The equilibrium results for the case with carbon formation for DRM process as a function of temperature are showed in Figure 7 at pressure of 1 bar (left side) and 10 bar (right side) by varying CO$_2$/CH$_4$ feed molar ratio, and in Figure 8 at CO$_2$/CH$_4$ feed molar ratio of 1 (left side) and 2 (right side) by varying pressure. At a relatively low temperature, the equilibrium constant of the Boudouard reaction is much higher than the unity and, thus, the formation of coke and CO$_2$ is favoured. This causes an enhanced conversion of methane due to a major presence of CO$_2$, whose consumption by RWGS is not sufficient to overcome the effect of the Boudouard reaction. At a moderately higher temperature, however, the influence of the Boudouard reaction is modest and, at the same time, that of RWGS increases. This causes a higher consumption of

CO_2, which tends to push the equilibrium of DRM towards the reactants leading to a slightly lower CH_4 conversion. On the other hand, at a higher temperature the DRM is favoured over both RWGS and Boudouard reactions, causing a boost in the CH_4 conversion. The overall result of these different tendencies is the presence of minima in the trends shown in Figures 7 and 8, which are justly caused by considering the effect of the Boudouard reaction.

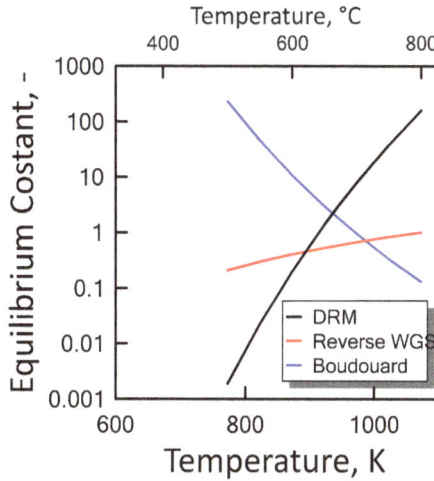

Figure 6. Equilibrium constant as a function of temperature for DRM (—), reverse WGS (—) and Boudouard (—) reactions.

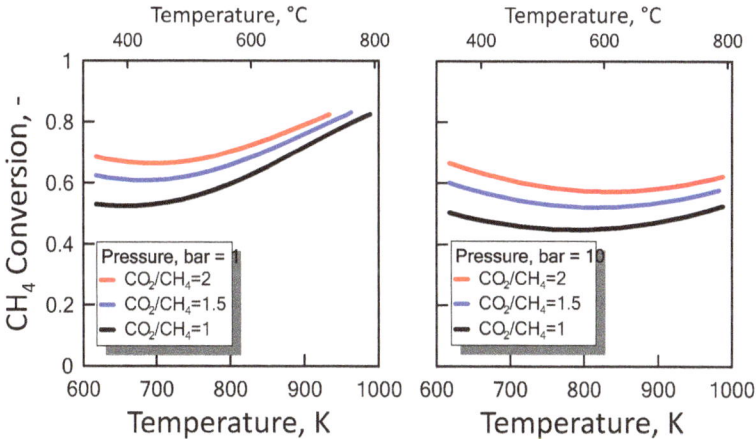

Figure 7. CH_4 equilibrium conversion at 1 bar (**left** side) and 10 bar (**right** side) as a function of temperature at different CO_2/CH_4 feed molar ratio taking into account carbon formation.

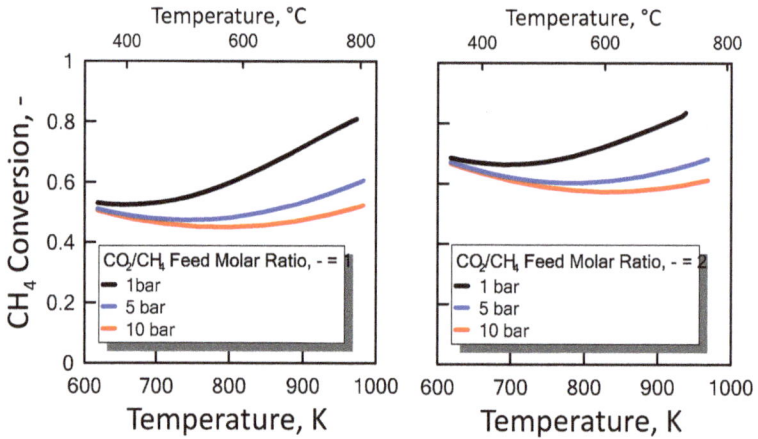

Figure 8. CH$_4$ equilibrium conversion at CO$_2$/CH$_4$ feed molar ratio of 1 (**left** side) and 2 (**right** side) as a function of temperature at different pressures taking into account carbon formation.

Furthermore, carbon formation is more important as both pressure and CO$_2$/CH$_4$ feed molar ratio increase.

Figure 9 shows the CH$_4$ equilibrium conversion as a function of pressure at temperature of 500 °C and a CO$_2$/CH$_4$ feed molar ratio of 1 for both cases previous described. In an ideal situation the optimum condition in which it operates is between the two equilibrium curves, i.e., lower than methane conversion which takes into account coke formation, and greater than methane conversion that does not take into account coke formation.

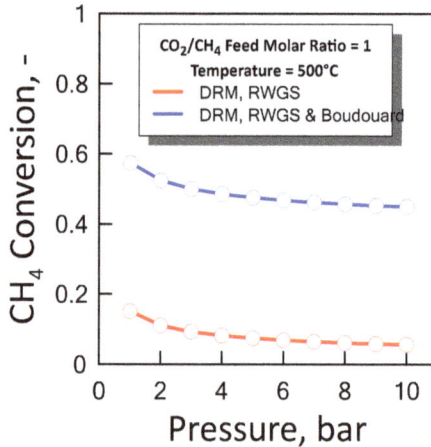

Figure 9. CH$_4$ equilibrium conversion at CO$_2$/CH$_4$ feed molar ratio of 1 and temperature of 500 °C as a function of pressure.

3.2. Permeation Measurements

Figure 10 shows the hydrogen permeating flux as a function of the Sieverts' permeation driving force at different values of temperature. As expected, hydrogen flux increases linearly with increasing driving force, indicating an ideal membrane behavior (i.e., permeation completely controlled by internal diffusion and hydrogen concentration in the metal lattice close to the infinite dilution).

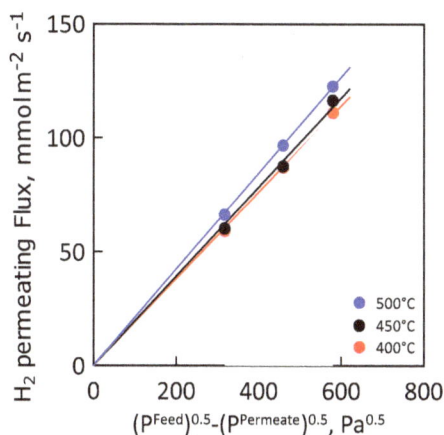

Figure 10. Hydrogen flux as a function of Sieverts' driving force at different temperatures.

This situation is confirmed by the Arrhenius-type plot of the hydrogen permeance, which shows a linear trend indicating an ideal permeance (and, thus, permeability, Figure 11). Table 4 summarizes the values of apparent activation energy, pre-exponential and hydrogen permeance at the temperatures considered.

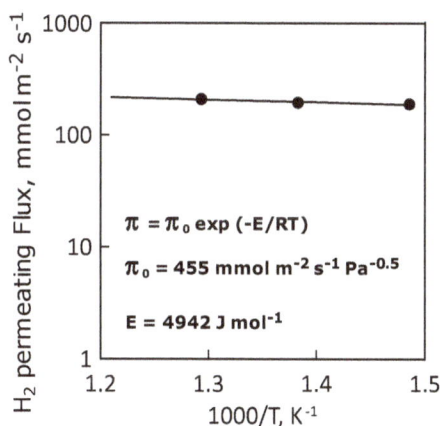

Figure 11. Arrhenius plot of hydrogen permeance.

Table 4. Apparent activation energy, pre-exponential factor and hydrogen permeance values.

Apparent Activation Energy, E	$J \cdot mol^{-1}$	4942
Permeance Pre-Exponential Factor	$mol \cdot m^{-2} \cdot s^{-1} \cdot Pa^{-0.5}$	455
Hydrogen Permeance, $mol \cdot m^{-2} \cdot s^{-1} \cdot Pa^{-0.5}$	400 °C	190
	450 °C	196
	500 °C	211

It must be noticed that the reaction experiments lasted ca. 500 h—alternating reaction with regeneration by methanation—and no drop of separating properties of the membrane, measured before and after reaction, was observed during this time.

3.3. Reaction Measurements

Before performing the DRM reaction measurements in MR, the catalyst activation was carried out in the presence of gas mixture having a molar concentration 90:10 = H_2:N_2 at 400 °C and atmospheric pressure for two hours. The values of CH_4 and CO_2 feed flow rates were calculated and set by fixing CO_2/CH_4 feed molar ratio, GHSV (space velocity), temperature and feed pressure. Figure 12 (left axis) shows the CH_4 conversion as a function of the feed pressure at GHSV of 100 h^{-1}, temperature of 500 °C and CO_2/CH_4 feed molar ratio of 1.5. The error bars indicate the carbon balance. The experimental conversion values obtained in the MR were compared with the equilibrium conversion obtained in the traditional reactor. In particular, the experimental CH_4 conversion was found to decrease with increasing feed pressure according to Le Chatelier-Braun principle since there is a net increase in the number of moles. It is noticeable that at sufficiently low feed pressures (1 bar in the specific case), the experimental conversions in the MR were higher than equilibrium conversion of a TR since the removal of the H_2 product shifts the equilibrium to the right. The permeation supplies such a significant contribution to overcoming the equilibrium performance. The higher feed pressure hinders the conversion since the reaction occurs with an increase in molar number, thus the hydrogen partial pressure on the feed side is lower and, similarly, its permeation.

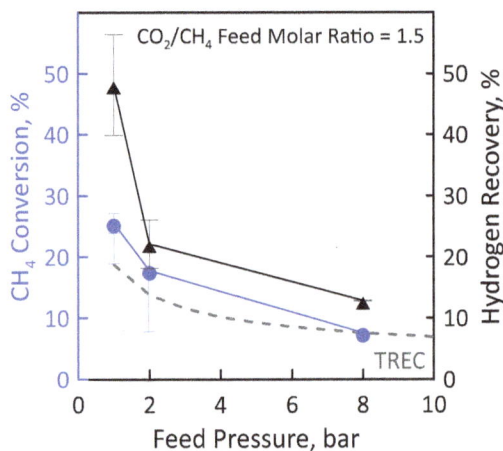

Figure 12. CH_4 conversion (●) and H_2 recovery (▲) as a function of feed pressures at CO_2/CH_4 feed molar ratio = 1.5, permeate pressure = 0.02 bar. Equilibrium conversion values in a traditional reactor (dashed line).

These data were obtained considering a permeate pressure of 0.02 bar. We decided to operate under vacuum for increasing the permeation rate. CH_4 conversion obtained at the permeate pressure of 1 bar is less than that of the equilibrium. With a low hydrogen permeate pressure value it is, thus, possible to significantly exceed the thermodynamic limit of the traditional reactor, reaching conversion significantly higher.

A higher methane conversion means a higher methane amount that reacts for producing hydrogen. For this reason, the hydrogen recovery follows the same trend as that of conversion (Figure 12, right axis). At a feed pressure of 1 bar and 0.02 bar of vacuum on the permeate, about 47% of hydrogen is recovered as a pure stream in the permeate. A higher feed pressure means a lower CH_4 conversion, and, thus, less hydrogen produced and that can be recovered. However, it has to be pointed out that, even in the worse conditions (i.e., 8 bar), around 15% of hydrogen is recovered.

Another reaction measurement was carried out at stoichiometric feed molar ratio under the same operating conditions as the just-analyzed ones (Figure 13). Additionally, in this case, the CH_4 conversion decreases with pressure (Figure 13, left axis). At low pressures, the experimental conversions in the MR are higher than equilibrium conversions since the removal of H_2 product shifts the equilibrium to the right. However, at the high pressure, the conversion in the MR is lower than the equilibrium one. This is a non-favourable condition because, despite using an MR, the H_2 permeation through the membrane cannot compensate the feed pressure negative effect induced by thermodynamics. Therefore, the reactor behavior is similar to that of a TR. Figure 13 (right axis) shows the effect of feed pressure on the H_2 recovery. Although conversion is low, a hydrogen recovery of about 20% was found. This could mean that, despite the small amount of methane converted, a sufficiently large amount of hydrogen is produced by reaction to permeate through the membrane. In an MR, a high feed pressure involves an increase of hydrogen permeation driving force, favouring a higher hydrogen removal from the reaction side towards the permeate side with a consequently higher hydrogen recovery. The opposite trend between methane conversion and hydrogen recovery can be justified by the combination of the negative effect induced by thermodynamics and the positive effect induced by permeation.

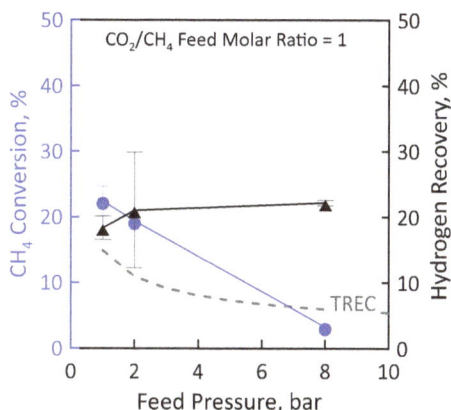

Figure 13. CH_4 conversion (●) and H_2 recovery (▲) as a function of feed pressures at CO_2/CH_4 feed molar ratio = 1, permeate pressure = 0.02 bar. Equilibrium conversion values in a traditional reactor (dashed line).

Figure 14 shows H_2/CO reaction selectivity as a function of the feed pressure for different feed molar ratio. As feed pressure increases, H_2/CO reaction selectivity decreases. It is less than 100% for all feed pressure range considered, that is, the CO amount produced is greater than that of H_2. This could mean that, operating at CO_2/CH_4 feed molar ratio of 1.5, H_2 product reacts with CO_2 fed (in excess) to produce CO and H_2O by reverse WGS side reaction.

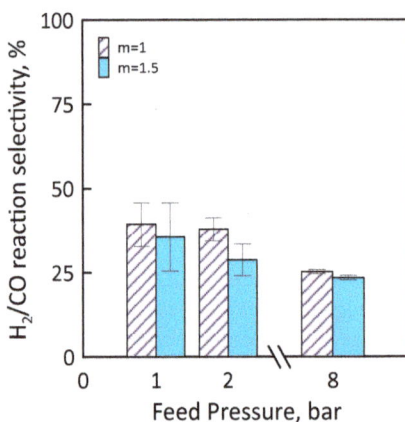

Figure 14. H_2/CO reaction selectivity as a function of feed pressures at CO_2/CH_4 feed molar ratio = 1–1.5, permeate pressure = 0.02 bar.

4. Conclusions

This work consists of an analysis of dry reforming of methane in a catalytic Pd-based membrane reactor. A 0.5 wt % Ru/Al_2O_3 catalyst was packed in the annulus between the shell and a Pd-Ag commercial membrane and the performance of the MR were analyzed as a function of feed pressure and feed molar ratio. The whole experimental period lasted about 500 h and any significant change in membrane permeation properties was observed during this time. Moreover, the catalyst was stable for the whole reaction period (alternated by a periodic regeneration with a diluted stream of 10% H_2 in argon), with no drops of its activity. Coke formation was continuously monitored, and carbon balance was below $\pm 10\%$. The higher conversion was achieved at low feed pressures owing to the favourable thermodynamics, reaching a higher value than the equilibrium one obtainable in a traditional reactor thanks to the hydrogen permeation through membrane. At 500 °C, MR showed good performance in terms of both CH_4 conversion and hydrogen recovery (CH_4 conversion = 26% and H_2 recovery = 46% @ 100 h^{-1}, 1 bar, 0.02 bar on permeate side) exceeding the traditional reactor equilibrium conversion. Comparing the results, it can be deduced that at a CO_2/CH_4 feed molar ratio of 1.5 there is a higher methane conversion but a lower H_2 recovery and H_2/CO reaction selectivity with respect to the stoichiometric feed molar ratio. This could indicate that both the reverse WGS and Boudouard reactions occur along with DRM, meaning that part of the H_2 produced is consumed to make water as a by-product, and part of CO produced is consumed to produce coke. So far, the main limitation to DRM industrialization remains as coke formation. The development of new catalysts with high and stable activity, is thus highly recommended. Membrane engineering with MR technology can play a fundamental role in the integration of the units and processes and, at the same time, in the definition of the knowledge necessary to drive the process by maximizing the gains both in terms of efficiency, productivity and plant size reduction.

Author Contributions: Conceptualization, A.B., G.B. and A.C.; Methodology, A.C.; Software, A.C.; Validation, M.G.; Formal Analysis, M.G. and A.C.; Investigation, M.G. and A.C. Resources, G.B. and A.C.; D.C., A.C.; Writing—Original Draft Preparation, A.C. and A.B.; Writing—Review & Editing, A.B. and A.C.; Supervision, A.B., A.C., G.B.

Funding: This research received no external funding.

Conflicts of Interest: The authors declare no conflict of interest.

Appendix A

Appendix A.1 Calculation Approach for Thermodynamic Equilibrium

The following sub-sections report the details of the calculation approach used to evaluate the equilibrium trend of DRM in the presence of the following side-reactions: The RWGS and Boudouard reaction.

The total Gibbs free energy of a single-phase only for a gas or a solid system can be respectively represented by Equations (A1) and (A2):

$$G^{tg}_{(T,P)} = \sum_{i=1}^{N} n_i G_i^{°g} + RT \sum_{i=1}^{N} n_i \ln \left(\frac{f_i^g}{f_i^{°g}} \right) \tag{A1}$$

$$G^{ts}_{(T,P)} = \sum_{i=1}^{N} n_i G_i^{°s} + RT \sum_{i=1}^{N} n_i \ln \left(\frac{f_i^s}{f_i^{°s}} \right) \tag{A2}$$

Hence, the total Gibbs free energy of a two-phase system is written as Equation (A3):

$$G^{t}_{(T,P)} = \sum_{i=1}^{N} n_i G_i^{°g} + RT \sum_{i=1}^{N} n_i \ln \left(\frac{f_i^g}{f_i^{°g}} \right) + \sum_{i=1}^{N} n_i G_i^{°s} + RT \sum_{i=1}^{N} n_i \ln \left(\frac{f_i^s}{f_i^{°s}} \right) \tag{A3}$$

The standard state is defined as the pure ideal gas state at 1 atm, $f_i^{°g} = P^{°} = 1\ atm$ and $G_i^{°g}$ equals to zero for each element in its standard state. The solid phase is assumed as a pure solid carbon and its reference state is at atmospheric pressure and 25 °C. The partial fugacity is written as shown in Equation (A4):

$$f_i^s = f_i^{°s} \tag{A4}$$

The total Gibbs free energy of the system is expressed in Equation (A5) by the summation over N species:

$$G^t = \sum_{i=1}^{N} n_i G_i^{°} + RT \sum_{i=1}^{N} n_i \ln \frac{f_i}{f_i^{°}} + n_s G_s \tag{A5}$$

where $G_i^{°}$ is the Gibbs free energy of species i under standard conditions, R is the universal molar gas constant, $f_i^{°}$ and f_i are fugacity of species i at standard and operating conditions, respectively, n_i is the number of moles of species i, and T is the temperature. If carbon is formed in the overall reactions, n_s represents the number of solid carbon molecules, while G^s is the Gibbs free energy of solid carbon at the operating conditions.

To reach equilibrium, Gibbs energy is minimized with respect to reaction degree ξ for which there are constraints to be respected (Equation (A6)). The necessary condition to have a minimum of G^t is reported in Equation (A7).

$$\min_{\xi_1, \xi_2, \xi_3} G^t_{T,P}(\xi)$$

$$s.t.\ 0 \le \xi_1 \le 1 \tag{A6}$$
$$0 \le \xi_2 \le 1$$
$$0 \le \xi_3 \le 1$$

$$\frac{dG^t}{d\xi} = 0 \tag{A7}$$

The vector of the reaction degrees ξ is written between the species involved in a reaction by a mass balance between both the initial and generic condition (Equation (A8)). As the number species in the system, more variables are necessary. In particular, many reactions degree as reactions are in the system. A reaction constrains the moles number variation of each species through the stoichiometry. Moles number must be positive.

$$N_i = N_i^0 + \sum_{j=1}^{N_R} v_{ij}\, \xi_j \tag{A8}$$

where N_i^0 is moles number of the specie i at reaction inlet (=0), v_{ij} is stoichiometric coefficient of specie i in the reaction j and N_R is number of independent reactions.

Appendix A.2 Equilibrium Calculation: Numerical Procedure

The thermodynamic equilibrium condition of a reaction system is calculated using the Gibbs free energy minimization method, at both constant P and T and with given initial composition. This method is based on the principle that the total Gibbs energy of system has its minimum value at chemical equilibrium. Furthermore, it requires the formalization of the reactions and the identification of all the species, beyond the reactants, which may be present at equilibrium.

The reactions used must be linearly independent, that is the stoichiometric matrix rank v (whose elements v_{ij} correspond to the stoichiometric coefficient of species i in the reaction j) must be equal to the independent reactions number (Equation (A9)). The latter variable is obtained by Equation (A10).

$$Rank(v) = Number\ of\ Indipendent\ Reaction \tag{A9}$$

$$Number\ of\ Indipendent\ Reaction = C - Rank(A) \tag{A10}$$

where C is a species number and A is an atom/species matrix.

The independent reactions systems considered in this paper work, for dry reforming of methane process, are: DRM (Equation (1)), reverse WGS (Equation (2)) and Boudouard (Equation (3)) reactions.

MATLAB was used in the equilibrium problem resolution. In particular, *fmincon* (MATLAB solver) attempts to find a constrained minimum of a scalar function of several variables starting at an initial estimate. This is generally referred to as constrained nonlinear optimization. During the process of optimization, the equilibrium condition is evaluated by varying the temperature (600:5:1073 K), pressure (1:1:10 bar) and molar feed ratio CO_2/CH_4 (1:0.5:2) value, whereas catalyst, reaction kinetics, and the transport process are not considered.

References

1. Henriques, I.; Sadorsky, P. Investor implications of divesting from fossil fuels. *Glob. Financ. J.* **2017**, in press. [CrossRef]
2. Hanley, E.S.; Deane, J.P.; Gallachóir, B.P.Ó. The role of hydrogen in low carbon energy futures—A review of existing perspectives. *Renew. Sustain. Energy Rev.* **2017**, *82*, 3027–3045. [CrossRef]
3. Hydrogen: Fuel for Our Future? Available online: http://www.worldwatch.org/node/4516 (accessed on 9 October 2018).
4. Wadhwani, S.; Wadhwani, A.K.; Agarwal, R.B. Clean Coal Technologies—Recent Advances. In Proceedings of the First International Conference on Clean Coal Technologies for Our Future, Chia Laguna, Sardinia, Italy, 21–23 October 2002.
5. Muradov, N. Low to near-zero CO_2 production of hydrogen from fossil fuels: Status and perspectives. *Int. J. Hydrogen Energy* **2017**, *42*, 14058–14088. [CrossRef]
6. Vivas, F.J.; de las Heras, A.; Segura, F.; Andújar, J.M. A review of energy management strategies for renewable hybrid energy systems with hydrogen backup. *Renew. Sustain. Energy Rev.* **2018**, *82*, 126–155. [CrossRef]
7. Akhtera, P.; Farkhondehfal, M.A.; Hernández, S.; Hussain, M.; Fina, A.; Saracco, G.; Khan, A.U.; Russo, N. Environmental issues regarding CO_2 and recent strategies for alternative fuels through photocatalytic reduction with titania-based materials. *J. Environ. Chem. Eng.* **2016**, *4*, 3934–3953. [CrossRef]

8. Pomilla, F.R.; Brunetti, A.; Marcì, G.; García-López, E.I.; Fontananova, E.; Palmisano, L.; Barbieri, G. CO_2 to liquid fuels: Photocatalytic conversion in a continuous membrane reactor. *ACS Sustain. Chem. Eng.* **2018**, *6*, 8743. [CrossRef]

9. Sellaro, M.; Bellardita, M.; Brunetti, A.; Fontananova, E.; Palmisano, L.; Drioli, E.; Barbieri, G. CO_2 conversion in a photocatalytic continuous membrane reactor. *RSC Adv.* **2016**, *6*, 67418–67427. [CrossRef]

10. Kohen, A.; Cannio, R.; Bartolucci, S.; Klinman, J.P. Enzyme dynamics and hydrogen tunneling in a thermophilic alcohol dehydrogenase. *Nature* **1999**, *399*, 496–499. [CrossRef] [PubMed]

11. Marpani, F.; Pinelo, M.; Meyer, A.S. Enzymatic conversion of CO_2 to CH_3OH via reverse dehydrogenase cascade biocatalysis: Quantitative comparison of efficiencies of immobilized enzyme systems. *Biochem. Eng. J.* **2017**, *127*, 217–228. [CrossRef]

12. Nocera, D.G. The artificial leaf. *Acc. Chem. Res.* **2012**, *45*, 767–776. [CrossRef] [PubMed]

13. Bensaid, S.; Centi, G.; Garrone, E.; Perathoner, S.; Saracco, G. Towards artificial leaves for solar hydrogen and fuels from carbon dioxide. *ChemSusChem* **2012**, *5*, 500–521. [CrossRef] [PubMed]

14. Chabi, S.; Papadantonakis, K.M.; Lewis, N.S.; Freund, M.S. Membranes for artificial photosynthesis. *Energy Environ. Sci.* **2017**, *10*, 1320–1338. [CrossRef]

15. Guczi, L.; Stefler, G.; Geszti, O.; Sajó, I.; Pászti, Z.; Tompos, A.; Schay, Z. Methane dry reforming with CO_2: A study on surface carbon species. *Appl. Catal. A Gen.* **2010**, *375*, 236–241. [CrossRef]

16. Bucharkina, T.V.; Gavrilova, N.N.; Kryzhanovskiy, A.S.; Skudin, V.V.; Shulmin, D.A. Dry reforming of methane in contactor and distributor membrane reactors. *Pet. Chem.* **2015**, *55*, 932–939. [CrossRef]

17. Usman, M.; Wan Daud, W.M.A.; Abbas, H.F. Dry reforming of methane: Influence of process parameters—A review. *Renew. Sustain. Energy Rev.* **2015**, *45*, 710–744. [CrossRef]

18. Kim, S.; Ryi, S.K.; Lim, H. Techno-economic analysis (TEA) for CO_2 reforming of methane in a membrane reactor for simultaneous CO_2 utilization and ultra-pure H_2 production. *Int. J. Hydrogen Energy* **2018**, *43*, 5881–5893. [CrossRef]

19. Roy, P.S.; Song, J.; Kim, K.; Park, C.S.; Raju, A.S.K. CO_2 conversion to syngas through the steam-biogas reforming process. *J. CO_2 Util.* **2018**, *25*, 275–282. [CrossRef]

20. Oyama, S.T.; Hacarlioglu, P.; Gu, Y.; Lee, D. Dry reforming of methane has no future for hydrogen production: Comparison with steam reforming at high pressure in standard and membrane reactors. *Int. J. Hydrogen Energy* **2012**, *37*, 10444–10450. [CrossRef]

21. Drioli, E.; Barbieri, G.; Brunetti, A. *Membrane Engineering for the Treatment of Gases*, 2nd ed.; Royal Society of Chemistry: Cambridge, UK, 2017; ISBN 978-1-78262-875-0.

22. Wang, S.; Lu, G.; Millar, G.J. Carbon dioxide reforming of methane to produce synthesis gas over metal-supported catalysts: State of the art. *Energy Fuel* **1996**, *10*, 896–904. [CrossRef]

23. Bosko, M.L.; Munera, J.F.; Lombardo, E.A.; Cornaglia, L.M. Dry reforming of methane in membrane reactors using Pd and Pd–Ag composite membranes on a NaA zeolite modified porous stainless steel support. *J. Membr. Sci.* **2010**, *364*, 17–26. [CrossRef]

24. Gallucci, F.; Tosti, S.; Basile, A. Pd-Ag tubular membrane reactors for methane dry reforming: A reactive method for CO_2 consumption and H_2 production. *J. Membr. Sci.* **2008**, *317*, 96–105. [CrossRef]

25. García-García, F.R.; Soria, M.A.; Mateos-Pedrero, C.; Guerrero-Ruiz, A.; Rodríguez-Ramos, I.; Li, K. Dry reforming of methane using Pd-based membrane reactors fabricated from different substrates. *J. Membr. Sci.* **2013**, *435*, 218–225. [CrossRef]

26. Silva, F.A.; Hori, C.E.; da Silva, A.M.; Mattos, L.V.; Munera, J.; Cornaglia, L. Hydrogen production through CO_2 reforming of CH_4 over Pt/$CeZrO_2$/Al_2O_3 catalysts using a Pd-Ag membrane reactor. *Catal. Today* **2012**, *193*, 64–73. [CrossRef]

27. Fedotov, A.S.; Antonov, D.O.; Uvarov, V.I.; Tsodikov, M.V. Original hybrid membrane-catalytic reactor for the Co-Production of syngas and ultrapure hydrogen in the processes of dry and steam reforming of methane, ethanol and DME. *Int. J. Hydrogen Energy* **2018**, *43*, 7046–7054. [CrossRef]

28. Sumrunronnasak, S.; Tantayanon, S.; Kiatgamolchai, S.; Sukonket, T. Improved hydrogen production from dry reforming reaction using a catalytic packed-bed membrane reactor with Ni-based catalyst and dense PdAgCu alloy membrane. *Int. J. Hydrogen Energy* **2016**, *41*, 2621–2630. [CrossRef]

29. Alique, D.; Martinez-Diaz, D.; Sanz, R.; Calles, J.A. Review of supported Pd-based membranes preparation by electroless plating for ultra-pure hydrogen production. *Membranes* **2018**, *8*, 5. [CrossRef] [PubMed]

30. Safariamin, M.; Tidahy, L.H.; Abi-Aad, E.; Siffert, S.; Aboukais, A. Dry reforming of methane in the presence of ruthenium-based catalysts. *C. R. Chim.* **2009**, *12*, 748–753. [CrossRef]

31. Jeon, J.; Nam, S.; Ko, C.H. Rapid evaluation of coke resistance in catalysts for methane reforming using low steam-to-carbon ratio. *Catal. Today* **2018**, *309*, 140–146. [CrossRef]

32. Simakov, D.S.A.; Leshkov, Y.R. Highly efficient methane reforming over a low-loading Ru/γ-Al_2O_3 catalyst in a Pd-Ag membrane reactor. *AIChE J.* **2018**, *64*, 3101–3108. [CrossRef]

33. Lee, D. Catalytic Reforming of CH_4 with CO_2 in a Membrane Reactor: A Study on Effect of Pressure. Doctoral Dissertations, VirginiaTech, Blacksburg, VA, USA, 2003; pp. 66–91.

34. Chein, R.Y.; Chen, Y.C.; Yu, C.T.; Chung, J.N. Thermodynamic analysis of dry reforming of CH_4 with CO_2 at high pressures. *J. Nat. Gas Sci. Eng.* **2015**, *26*, 617–629. [CrossRef]

chemengineering

MDPI

Article

Ionic Liquid Hydrogel Composite Membranes (IL-HCMs)

Shabnam Majidi Salehi [1,2], **Rosangela Santagada** [1], **Stefania Depietra** [1], **Enrica Fontananova** [1], **Efrem Curcio** [2] and **Gianluca Di Profio** [1,*]

[1] National Research Council of Italy (CNR), Institute on Membrane Technology (ITM), Via P. Bucci c/o Università della Calabria Cubo 17/C, 87036 Rende (CS), Italy; sh.majidi.s@gmail.com (S.M.S.); rosangela902@virgilio.it (R.S.); stefydep92@hotmail.it (S.D.); e.fontananova@itm.cnr.it (E.F.)
[2] University of Calabria (UNICAL), Department of Environmental and Chemical Engineering (DIATIC), Via P. Bucci Cubo 45/C, 87036 Rende (CS), Italy; e.curcio@unical.it
*   Correspondence: g.diprofio@itm.cnr.it; Tel.: +39 0984 492010; Fax: +39-0984-402103

Received: 19 February 2019; Accepted: 26 April 2019; Published: 5 May 2019

Abstract: In this work, novel hydrogel composites membranes comprising [2-(Methacryloyloxy)ethyl] dimethyl-(3-sulfopropyl)ammonium hydroxide as monomer, N,N-methylene bisacrylamide as cross-linker, and 1-butyl-3-methylimidazolium hexafluorophosphate as ionic liquid additive, have been developed. Ionic liquid hydrogel composite membranes (IL-HCMs) were tested for membrane contactors applications, aiming to reduce surface hydrophobicity of the polypropylene support, to reduce wetting tendency due to interaction with hydrophobic foulants, while affecting salts rejection in desalination operation, because of the entrapment of ILs inside the porous mesh-like structure of the gel layer. Transmembrane flux comparable to the sole polypropylene support was observed for IL content > 1 wt.%. Furthermore, all IL membranes presented a larger rejection to sodium chloride than the PP support or the composites without ionic liquid inside. Although the overall transmembrane flux of IL-HCMs developed in this work is comparable with that of state of the art MD membranes, this study demonstrated that the strong hydrophilic hydrogel layer, with C.A. < 50° for IL content larger than 1 wt.%, serves as a stabilization coating, by providing the new media between the feed and the hydrophobic membrane surface, thus potentially controlling the diffusion of hydrophobic foulant molecules. This would result in a decrease in the membrane wetting and fouling aptitude.

Keywords: advanced separations; desalination; hydrogel composite membranes; ionic liquids membranes; membrane distillation

---

1. Introduction

Hydrogels are soft materials consisting of a mesh-like structure with polymer chains physically or chemically cross-linked and water molecules filling the interstitial spaces [1,2]. Due to their permeable net-like consistence, hydrogels can exert separation functions based on molecular size exclusion [3]. Since the mesh size can be modulated by adapting the swelling state in response to externally applied stimuli, like temperature, pH, ionic strength, interaction with specific molecules, electric or magnetic field, transport of specific solution components through hydrogels can be controlled by a stimuli-responsive behavior [4–14].

Because of their unique features as separative media, hydrogels are facing notable implications in the field of membrane processes [15]. The combination of a hydrogel layer with a porous support provides favorable synergisms, resulting in composite materials with totally new separation functions with respect to the bulk hydrogel or the substrate, while affording improved mechanical stability of the supported soft gel phase.

Recently, we have developed several hydrogel composite membranes, containing a hydrophobic support and a hydrophilic surface layer, whose main application is in the field of membrane contactors. Hydrogel composite membranes (HCMs) have been developed: (1) for membrane distillation (MD) [16], with enhanced salt rejection under Donnan exclusion effect generated by the polyelectrolyte nature of the gel phase, (2) as protein crystallization environment [17,18], to produce crystals at lower supersaturation and of larger size compared to conventional crystallization methods, (3) as biomimetic mineralization platform [19], for the synthesis of calcium carbonate structures displaying multi-scale and hierarchical architecture, and (4) as antibacterial food package materials [20], due to the effect of enzyme crystals included in the hydrogel layer.

In this study, ionic liquid hydrogel composite membranes (IL-HCMs) have been prepared by layering a [2-(Methacryloyloxy)ethyl]dimethyl-(3-sulfopropyl)ammonium hydroxide hydrogel, cross-linked with N,N-methylene bisacrylamide in the presence of 1-butyl-3-methylimidazolium hexafluorophosphate as ionic liquid (IL) additive, and tested for MD applications. The main purpose of this work was the development of a suitable membrane functionalization strategy, that would be effective in reducing the hydrophobic interaction of the polymeric surface with potential foulants, thanks to the presence of the more hydrophilic hydrogel layer, while enforcing salts rejection due to the contribution of the ILs entrapped in the porous structure of the gel layer.

In fact, membrane fouling is particularly concerning when intrinsically hydrophobic membrane materials are used to treat feed solutions containing an abundance of organic components, due to their strong hydrophobic interactions with the polymeric surface [21–23]. This is particularly detrimental in MD applications, where severe fouling of the hydrophobic membrane is the starting point of pore wetting. In such cases, extensive feed pre-treatment is requested to remove the hydrophobic contaminants ahead the membrane process, with significant increases of treatment costs. Therefore, membrane surface modification is the most common method to improve membrane anti-fouling and anti-wetting properties in MD. For instance, shifting membrane surface hydrophobicity toward super hydrophobicity allows to generate an air gap between the liquid and the surface that helps to increase the admissible pore size prior to pore wetting occurrence, thus ensuring higher mass flux and membrane stability [24,25]. On the other hand, changing the membrane surface from hydrophobic to more hydrophilic behavior, imparts robust resistance to fouling [26–28], thanks to the formation of a hydration shell which renders the membrane surface less susceptible to interaction with the hydrophobic contaminants [29–32]. Such kind of membrane functionalization strategy is potentially suitable for membrane contactors applications, such as membrane distillation and membrane crystallization, since the hydrophilic layer protects membrane's surface from low-surface-tension components, thus preventing feed solutions from penetrating into the pores [33,34].

While the development of membranes with (super) hydrophilic skin layer has allowed the applications of MD to desalinate more challenging feed waters [22,35], little studies have been proposed so far on the development of hydrogel-functionalized membranes for membrane contactors. In a previous study, a thin layer of agarose hydrogel was positioned on the surface of a Teflon membrane, providing good anti-wetting behavior in the presence of surfactant in MD desalination [36]. In another study [37], composite membranes including poly(diallyldimethylammonium chloride)/poly acrylic acid semi-interpenetrating hydrogel on a polyvinylidene fluoride support, showed long-term robustness with 3 wt.% NaCl solution containing cationic, ionic and non-ionic surfactants in direct contact MD experiments. Here, we show that, in addition to potentially improved membrane resistance to wetting due to the protective hydrophilic hydrogel layer, larger rejection to sodium chloride than the PP membrane can be generated under the effect of ionic liquid entrapped in the polyelectrolyte layer, with the resulting increase in charge density in the gel network that rejects ions at the solution/hydrogel interface by the Donnan exclusion potential.

2. Experimental Section

2.1. Materials

Polypropylene (PP) flat membranes (Accurel PP 2E HF, nominal pore size 0.2 μm) were purchased from Membrana GmbH (Germany). [2-(Methacryloyloxy)ethyl]dimethyl-(3-sulfopropyl)ammonium hydroxide (monomer, SPE, cod. 537284), N,N-methylene bisacrylamide (cross-linker, MBA, cod. 146062), 2-hydroxy-2-methyl propiophenone (photoinitiator, cod. 1001451059), and 1-Butyl-3-methylimidazolium hexafluoro phosphate (ionic liquid, IL, cod. 18122) (Figure 1) were from Sigma-Aldrich (Italy). Sodium chloride (NaCl, cod. 131659.1211) was from Panreac (Italy). Methanol (CH_3OH, HPLC grade, cod. 20844.320) was from VWR (Italy). All chemicals were used without any further purification. Milli-Q water was used for all solutions and as condensing fluid in MD tests.

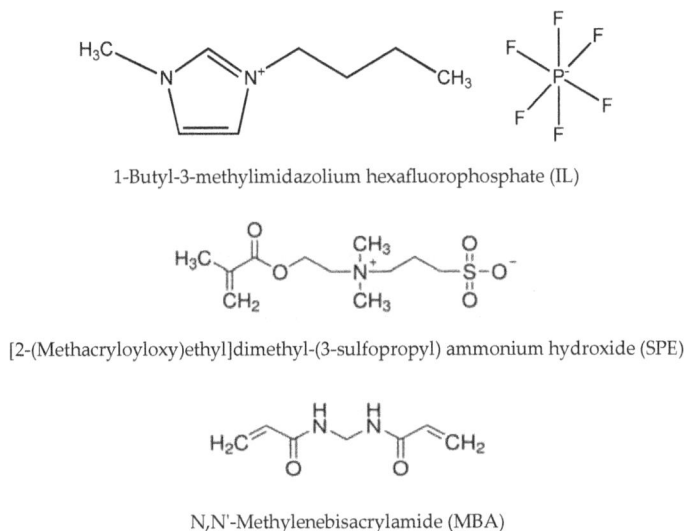

1-Butyl-3-methylimidazolium hexafluorophosphate (IL)

[2-(Methacryloyloxy)ethyl]dimethyl-(3-sulfopropyl) ammonium hydroxide (SPE)

N,N'-Methylenebisacrylamide (MBA)

Figure 1. Structural formula of hydrogel components used in this work.

2.2. Preparation of Hydrogel Composite Membranes

Polypropylene membranes, used as support for hydrogel composites, were conditioned by soaking in methanol for 24 h at room temperature and then dried with tissue paper immediately before use. The monomer SPE was dissolved in water by magnetic stirring (50 rpm) at 50 °C. The cross-linker MBA was then added to the monomer solution under stirring until complete dissolution. Subsequently, the photoinitiator was added to the solution together with the proper amount of IL additives (see Table 1). The solutions were then left to stir for 1 h at room temperature. The pre-polymer solutions were then cast onto the conditioned PP substrate by a micrometric film applicator (Elcometer 3570) at a selected thickness of the liquid film (100, 350, or 600 μm). Photo-initiated graft polymerization was then performed under the UV/Vis irradiation lamp (GR.E, 500 W) in a vented exposition chamber (Helios Italquartz, Italy), for 20 min. Composite membranes without IL were also prepared as reference samples. After that, composite membranes were washed and stored in water at room temperature for characterization and MD tests.

Table 1. Hydrogel composite membrane sample codes, compositions of the pre-polymerization solution and thickness of the casted film layer.

Membrane Sample Code	Monomer (wt.%)	Crosslinker (wt.%)	Photoinitiator (wt.%)	H₂O (wt.%)	IL (wt.%)	Casting Thickness (μm)
SPE-100	10.0	1.0	0.3	88.7	0	100
SPE-IL1%-100	10.0	1.0	0.3	88.6	0.1	100
SPE-IL5%-100	10.0	1.0	0.3	88.2	0.5	100
SPE-IL10%-100	10.0	1.0	0.3	87.7	1.0	100
SPE-IL15%-100	10.0	1.0	0.3	87.2	1.5	100
SPE-350	10.0	1.0	0.3	88.7	0	350
SPE-IL1%-350	10.0	1.0	0.3	88.7	0.1	350
SPE-IL5%-350	10.0	1.0	0.3	88.6	0.5	350
SPE-IL10%-350	10.0	1.0	0.3	88.2	1.0	350
SPE-IL15%-350	10.0	1.0	0.3	87.7	1.5	350
SPE-600	10.0	1.0	0.3	88.7	0	600
SPE-IL1%-600	10.0	1.0	0.3	88.7	0.1	600
SPE-IL5%-600	10.0	1.0	0.3	88.6	0.5	600
SPE-IL10%-600	10.0	1.0	0.3	88.2	1.0	600
SPE-IL15%-600	10.0	1.0	0.3	87.7	1.5	600

2.3. Membrane Morphology Examination

Morphological analyses of membranes top and cross section were performed by a Quanta 200 F FEI Philips scanning electron microscope (SEM). Samples were attached with carbon conductive tape to steal stubs and sputtered with chromium. The accelerating voltage was 12.0–15.0 kV under high vacuum. Samples were cryo-fractured with liquid nitrogen prior to examine membrane cross-section.

2.4. Water Contact Angle

Static contact angle was measured by a goniometer (Nordtest, Italy) at ambient temperature. A 5 μL drop of water was put onto the membrane sample by a micro-syringe and measurements were carried out by setting the tangents on both visible edges of the droplet. The average value of five measurements at different detection positions for each sample is reported in Table 2.

2.5. Chemical Surface Analysis

Surface chemical analysis of HCMs was carried out by Fourier transform infrared spectroscopy (Nicolet iS10 FT-IR spectrometer, Thermo Scientific, USA). Spectra were recorded in the range of 650–4000 cm^{-1} using an ATR (attenuated total reflectance) device, with a resolution of 1 cm^{-1}, and averaged over 30 scans.

2.6. Direct Contact Membrane Distillation Experiments

Figure 2 shows the direct contact membrane distillation (DCMD) equipment used in this work. Technical details are reported elsewhere [16]. The starting feed and condensing water volumes were 3 L each. The temperature of the feed and the distillate was 60 °C and 20 °C, respectively. Co-current solutions circulation flow rate was 12 L·h^{-1}. Active membrane was 3.5×10^{-3} m^2. Tests were performed with pure water or NaCl solution (1 g·L^{-1}) as feed. Each test lasted 6 h.

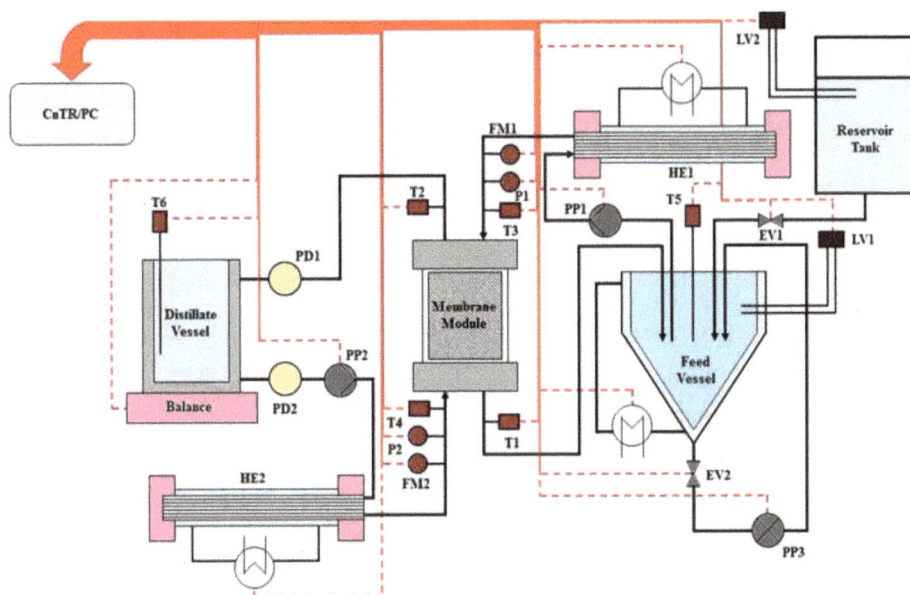

Figure 2. Membrane distillation plant used in direct contact configuration.

NaCl rejection was calculated by using Equation (1) by measuring the electrical conductivity (Jenway, Bibby Scientific, UK) of the distillate. Salt rejection R is defined as:

$$R = \left(1 - \frac{C_{distillate}}{C_{feed}}\right) \times 100 \tag{1}$$

where C_{feed} and $C_{distillate}$ are NaCl concentrations in the feed and in the distillate, respectively. R is quantified by mass balance after properly considering the electrical conductivity of the overall distillate with the time and the transmembrane flux under opportune calibration. Transmembrane flux J was taken as the average value detected under steady conditions (normally occurring during the last three hours of the test) and it is calculated as:

$$J = \frac{M}{\Delta t \cdot A} \tag{2}$$

with M the mass of liquid passed through the membrane in the time interval Δt, and A the effective membrane area.

3. Results and Discussion

Figure 3 shows an SEM cross-section image of an SPE-MBA ionic liquid hydrogel composite membrane (IL 10 wt.%) prepared by casting the pre-polymerization solution at 100-μm thickness. It is evident the homogeneous hydrogel layer, typically around 20 ± 5 μm thick, was well linked to the PP support. For the same pre-polymerization solution composition, changing the casing thickness of solution layer (100, 350, or 600 μm) resulted in different width of the final hydrogel layer upon polymerization under UV irradiation. Normally, 350 and 600 μm-thick liquid layers load on the support lead to hydrogel layers of 80 ± 20 μm and 150 ± 30 μm, respectively, after polymerization. No significant effect of the amount of IL on the final hydrogel layer thickness was observed during membrane preparation (variation less than 10%). In the case of IL > 10 wt.%, pre-polymerization solution presented some persistent phase separation upon mixing of the components.

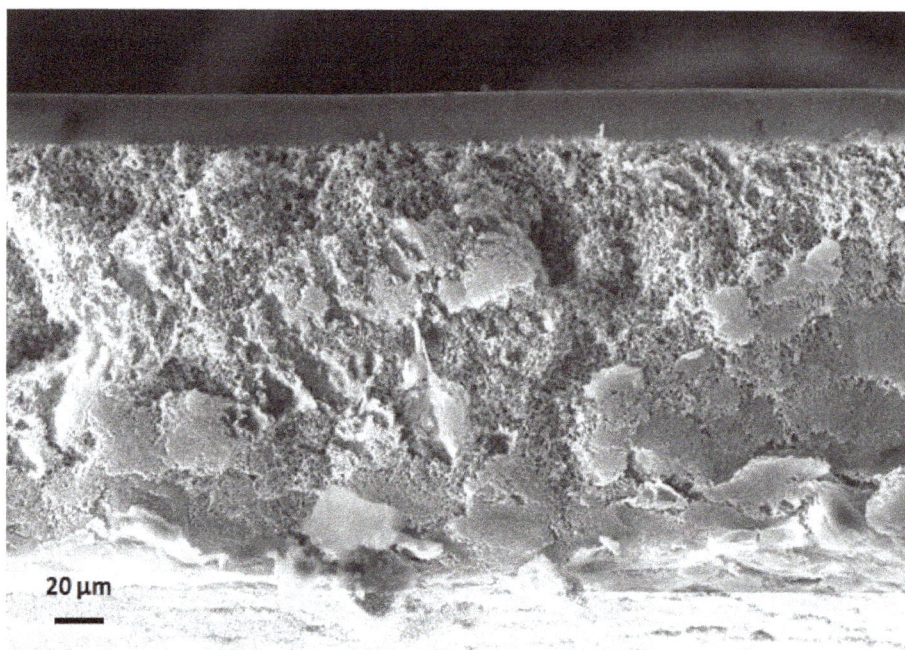

Figure 3. Typical scanning electron microscope (SEM) cross-section image of an SPE-MBA ionic liquid (IL 10 wt.%) hydrogel composite membrane (HCM) prepared by casting the pre-polymerization solution at 100-μm thickness.

Figure 4 displays ATR-FTIR spectra of IL-HCMs prepared with an increasing amount of IL. The growth of the peak around 840 cm^{-1}, assigned to the asymmetric stretching of the PF6 ion and bending of the imidazolium ring [38,39], with rising amounts of IL in the pre-polymerization solution can be observed. The strong electrostatic interactions between IL and SPE components provide stability of the hydrogel phase against the possible release of the IL from the composite membrane. The cationic part of the imidazolium group of the IL (Figure 1) interacts well with the sulfonic group of the SPE, while the ammonium part of SPE has an affinity for the hexafluorophosphate ion, thus keeping electroneutrality of the system.

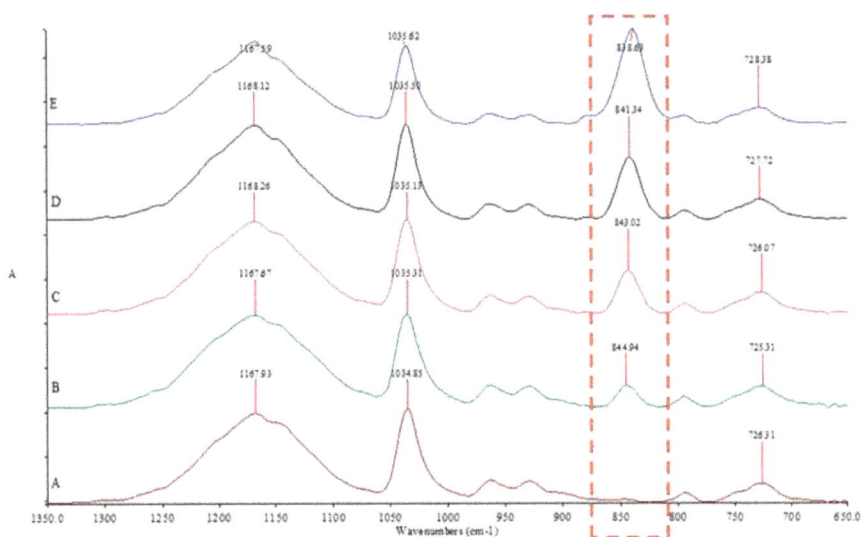

(a)

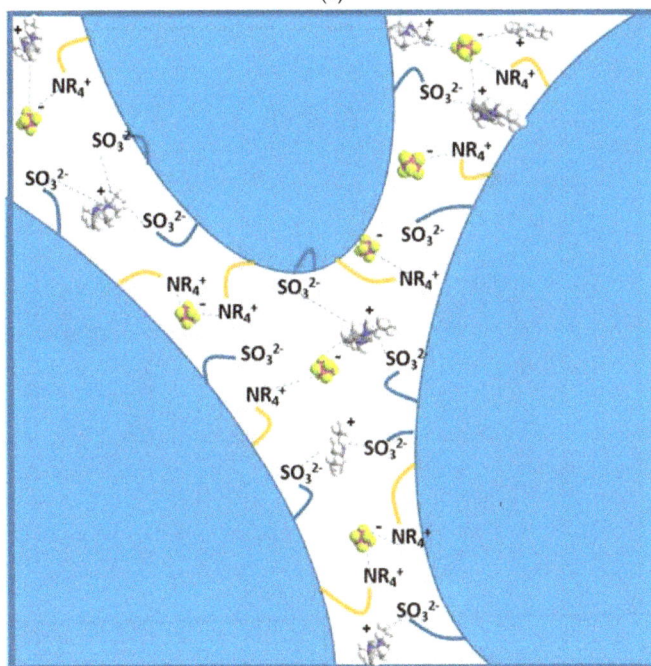

(b)

Figure 4. (**a**) ATR-FTIR spectra of ionic liquid hydrogel composite membranes (IL-HCMs) prepared with an increasing amount of IL (see Table 1): (A) SPE, (B) SPE-IL1%, (C) SPE-IL5%, (D) SPE-IL10%, (E) SPE-IL15%. (**b**) Schematic illustrating the effect of interaction between the fixed polyelectrolyte charges and mobile ionic liquid charges. White region denotes the hydrogel mesh structure with polymer chains and cross-linking points, blue spaces represent voids. The polymer mesh structure drawn here does not necessarily represent the effective system. The size of the ions relative to the mesh size is not drawn to scale.

Table 2 reports the water contact angles for IL-HCMs prepared at 350-μm thickness of the casting solution. As observed, the strong hydrophobic nature of the polypropylene support (138°) turns into hydrophilic (C.A. ~68°) with the functional hydrogel layered on the PP surface, while it becomes even more hydrophilic (C.A. < 50°) when introducing IL > 1 wt.% in the hydrogel layer.

Table 2. Water contact angles for IL-HCMs prepared at 350-μm film solution with different amounts of IL.

Membrane Sample Code	Water Contact Angle (°)
PP support	138.0 ± 1.0
SPE-350	67.9 ± 3.4
SPE-IL1%-350	61.1 ± 0.9
SPE-IL5%-350	45.6 ± 1.9
SPE-IL10%-350	44.5 ± 2.0
SPE-IL15%-350	40.8 ± 1.0

Composite membranes were tested for membrane distillation application by using the experimental equipment of Figure 2 with pure water or NaCl solution at 1 g·L^{-1} as feed. Figure 5 displays observed transmembrane fluxes *J* and salt rejections *R* performances. Despite the change of the nature of the membrane layer facing the feed from strongly hydrophobic to hydrophilic, all composite membranes demonstrated salt rejection > 99% over 6 h operation, indicating no occurrence of wetting, thus making them suitable for MD applications.

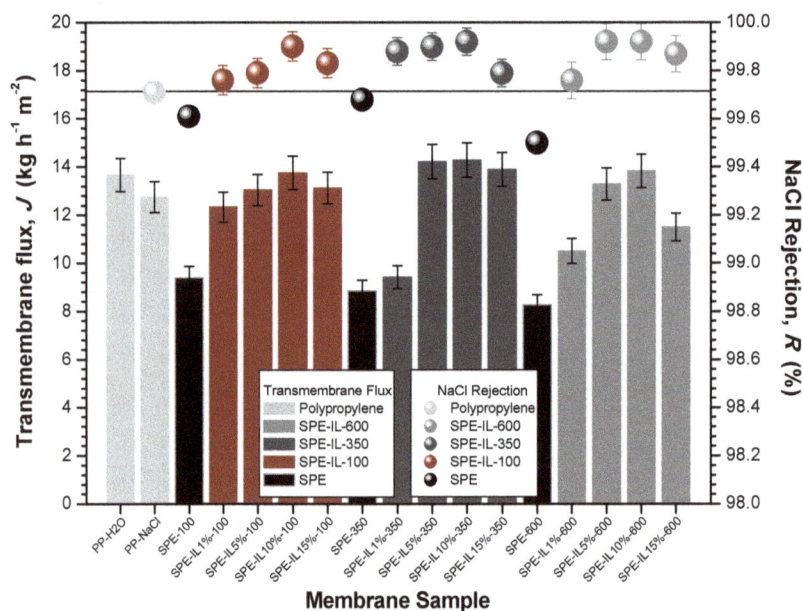

Figure 5. Observed transmembrane fluxes *J* and NaCl rejections *R* for tested membranes. Horizontal line is a guide for readers and refers to the rejection of NaCl salt of unfunctionalized PP support.

The first two bars of Figure 5 display the flux observed for the PP support alone with pure water (PP-H$_2$O) and with the saline solution (PP-NaCl) as feed, respectively. As noted, the slight decline in transmembrane flux (around to 7%) for NaCl feed solution compared to pure water, is due to the reduction of water vapour pressure under the effect of the dissolved salt, that decreases the driving force Δp (the vapour pressure gradient evaluated at the membrane surfaces) for mass-transfer in MD [40]. The observed rejection of NaCl for PP membrane support is 99.7%.

The consistent reduction of J is observed in Figure 5 for the membrane SPE-100, dropping by almost 30% (from 12.8 Kg·h^{-1}·m^{-2} for pristine PP to 9.4 Kg·h^{-1}·m^{-2}), when the hydrogel phase is layered on the surface of the PP support in the resulting HCM. This can be explained by the additional resistance to mass transfer, compared to untreated PP membranes, generated by the presence of the hydrogel. In the case of HCMs prepared by casting the pre-polymerization solution at 350 μm (SPE-350) and 600 μm (SPE-600), J was observed to scale almost linearly with the thickness of the casted solution. At 600 μm thickness, the reduction in J is almost 40% compared to the transmembrane flux registered for PP alone, dropping to 7.7 Kg·h^{-1}·m^{-2}.

When including IL at 1 wt.% in the casting solution (SPE-IL1%-100), the transmembrane flux of the hydrogel-coated membrane is almost completely recovered (only 3% less) with respect to the PP, while it grows regularly up to 13.8 Kg·h^{-1}·m^{-2} (i.e., larger than for the sole PP support) by increasing the amount of IL up to 10 wt.% (SPE-IL10%-100). In the latter case, rejection of NaCl reaches 99.99%.

Compared to a previous study, where transmembrane flux reduced by 71% when an agarose hydrogel layer was attached on the surface of a Teflon membrane [36], we found that including the IL additive in the hydrogel layer of HCMs allows almost complete salt rejection without affecting productivity.

The similar increasing trend of J is observed for HCMs prepared at 350 and 600 μm of casting solution, although the beneficial contribution of the IL to J at lower content (1 wt.%) is less effective than in the former case. Interestingly, increasing the amount of ionic liquid in the pre-polymerization solution, salt rejection increases regularly for all samples. For all the IL-HCMs, NaCl rejection overruns the observed value for the pristine PP support. Regardless of the gel layer thickness, the increase of ILs > 10 wt.% is associated with a slight decrease of both transmembrane flux and salt rejection. This behavior, in combination with the visual observation of liquid-liquid phase separation occurring in the pre-polymerization solution for the largest amount of ILs (15 wt.%), indicates that this solution composition is unsuitable for the preparation of IL-HCMs.

The singular response of IL-HCMs in MD testing, depending on the gel layer composition, is due to the synergistic interaction of the hydrogel phase and the ionic liquid dispersed in the mesh-like structure. According to experiments, it is clear that the presence of mobile ionic species in the polyelectrolyte gel network is responsible for the increased NaCl rejection. Figure 5, in fact, reveals that salt rejection for composite membranes without IL is always lower than that observed for the PP support. In the case of NaCl as a feed solution, polarized groups existing in the gel network affect mobile ions distribution at the hydrogel/feed interface by interacting with Na$^+$ cations and Cl$^-$ anions. The overall flux of ions (both Na$^+$ and Cl$^-$) from the bulk solution to the hydrogel phase is sustained by the diffusive transport under concentration gradient, which is also affected by the osmotic pressure generated by the attraction of water molecules by the strongly hydrophilic gel phase. In the absence of IL, the enhanced mobile charges density in the hydrogel network is responsible for the lower salt rejection. On the other hand, the increase of salt rejection observed for the membranes doped with IL can be rationalized considering the electrostatic repulsion (Donnan effect [41–43]) of either cationic or anionic species with charged moieties IL. Obviously, the mechanism of the Donnan exclusion in the case of ILs is more complex than simpler charged species, because of their unique molecular structure, polarity, and charge distribution [44]. SPE contains quaternary amminic and sulphonic groups (Figure 1) that may interact strongly with both cation and anion part of the IL. It is possible to speculate that the hydrophilic micro-channels formed by the interactions between the polyelectrolyte gel network and the IL (Figure 4) exert higher resistance to the transport of ionic species: Na$^+$ cations are rejected by the imidazolium component of the IL and Cl$^-$ anions are rejected by the phosphate ion. At the same time, the resistance to the water transport is reduced with respect to a hydrogel layer without IL, thanks to the well-interconnected network of hydrogen bonds formed between the two components (IL and polyelectrolyte).

4. Conclusions

In this work, ionic liquid hydrogel composite membranes were successfully prepared and tested for membrane contactors applications. Namely, [2-(Methacryloyloxy)ethyl]dimethyl-(3-sulfopropyl)ammonium hydroxide hydrogel, cross-linked with N,N-methylene bisacrylamide, and containing 1-butyl-3-methylimidazolium hexafluorophosphate as ionic liquid additive, have been demonstrated suitable for membrane distillation applications. Overall, water transport through IL-HCMs is not negatively affected by the presence of the hydrogel when the IL is included in the gel network. Complete recovery, and even improvement, of the transmembrane flux, compared to the sole hydrophobic support, was observed for IL content > 1 wt.%. Furthermore, all IL membranes presented a larger rejection of sodium chloride than the PP membrane or the composite without ionic liquid inside. This is due to the effect of the ionic liquid entrapped inside the polyelectrolyte layer with the resulting increase in charge density in the gel network that rejects ions at the feed/hydrogel interface under the effect of the Donnan exclusion potential. This effect brings salt rejection as high as 99.99% for composite membranes containing IL up to 10 wt.%. Increasing further the amount of IL induces phase separation in the pre-polymerization solution, thus making this composition unsuitable for IL-HCMs preparation.

Although the overall transmembrane flux of IL-HCMs developed in this work is comparable with that of state-of-the-art MD membranes, this study demonstrated that a hydrophobic macroporous membrane could be functionalized by a hydrogel layer comprising IL. The strong hydrophilic hydrogel, in addition to the increased transmembrane flux and salt rejection, serves as a stabilization layer, by providing the new media between the feed and the hydrophobic membrane surface, thus potentially controlling the diffusion of hydrophobic foulant molecules. This would result in a decrease in the membrane wetting and fouling aptitude.

Author Contributions: Conceptualization, G.D.P.; methodology, G.D.P., E.F., E.C.; investigation, S.M.S., R.S., and S.D.; data curation, S.M.S., R.S., and S.D.; writing-original draft preparation, G.D.P.; writing-review and editing, G.D.P., E.F., and E.C.; supervision, G.D.P., E.F.; funding acquisition, G.D.P., E.C.

Funding: The Education, Audiovisual and Culture Executive Agency (EACEA) for the financial support of the doctoral research fellowship of S. M. S. through the programme Erasmus Mundus Doctorate in Membrane Engineering—EUDIME (FPA 2011-0014).

Acknowledgments: Authors wish to thank The Education, Audiovisual and Culture Executive Agency (EACEA) for the financial support to the doctoral research fellowship of S. M. S. through the programme Erasmus Mundus Doctorate in Membrane Engineering—EUDIME (FPA 2011-0014).

Conflicts of Interest: The authors declare no conflict of interest.

References

1. Osada, Y.; Gong, J.P. Soft and Wet Materials: Polymer Gels. *Adv. Mater.* **1998**, *10*, 827. [CrossRef]
2. Peppas, N.A.; Hilt, J.Z.; Khademhosseini, A.; Langer, R. Hydrogels in Biology and Medicine: From Molecular Principles to Bionanotechnology. *Adv. Mater.* **2006**, *18*, 1345–1360. [CrossRef]
3. Tong, J.; Anderson, J.L. Partitioning and diffusion of proteins and linear polymers in polyacrylamide gels. *Biophys. J.* **1996**, *70*, 1505–1513. [CrossRef]
4. Tenhaeff, W.E.; Gleason, K.K. Surface-Tethered pH-Responsive Hydrogel Thin Films as Size-Selective Layers on Nanoporous Asymmetric Membranes. *Chem. Mater.* **2009**, *21*, 4323–4331. [CrossRef]
5. Hirokawa, Y.; Tanaka, T. Volume phase transition in a nonionic gel. *J. Chem. Phys.* **1984**, *81*, 6379–6380. [CrossRef]
6. Polotsky, A.A.; Plamper, F.A.; Borisov, O.V. Collapse-to-Swelling Transitions in pH- and Thermoresponsive Microgels in Aqueous Dispersions: The Thermodynamic Theory. *Macromolecules* **2013**, *46*, 8702–8709. [CrossRef]
7. Beebe, D.J.; Moore, J.S.; Bauer, J.M.; Yu, Q.; Liu, R.H.; Devadoss, C.; Jo, B.-H. Functional hydrogel structures for autonomous flow control inside microfluidic channels. *Nature* **2000**, *404*, 588–590. [CrossRef]

8. Chang, C.; He, M.; Zhou, J.; Zhang, L. Swelling Behaviors of pH- and Salt-Responsive Cellulose-Based Hydrogels. *Macromolecules* **2011**, *44*, 1642–1648. [CrossRef]
9. Tanaka, T.; Nishio, I.; Sun, S.T.; Uenonishio, S. Collapse of Gels in an Electric Field. *Science* **1982**, *218*, 467–469. [CrossRef]
10. Suzuki, A.; Tanaka, T. Phase transition in polymer gels induced by visible light. *Nature* **1990**, *346*, 345–347. [CrossRef]
11. Mosiewicz, K.A.; Kolb, L.; van der Vlies, A.J.; Martino, M.M.; Lienemann, P.S.; Hubbell, J.A.; Ehrbar, M.; Lutolf, M.P. In situ cell manipulation through enzymatic hydrogel photopatterning. *Nat. Mater.* **2013**, *12*, 1072–1078. [CrossRef]
12. Frey, W.; Meyer, D.E.; Chilkoti, A. Dynamic addressing of a surface pattern by a stimuli-responsive fusion protein. *Adv. Mater.* **2003**, *15*, 248–251. [CrossRef]
13. Mart, R.J.; Osborne, R.D.; Stevens, M.M.; Ulijn, R.V. Peptide-based stimuli-responsive biomaterials. *Soft Matter* **2006**, *2*, 822–835. [CrossRef]
14. Wilson, A.N.; Guiseppi-Elie, A. Bioresponsive Hydrogels. *Adv. Healthcare Mater.* **2013**, *2*, 520–532. [CrossRef] [PubMed]
15. Yang, Q.; Adrus, N.; Tomicki, F.; Ulbricht, M. Composites of functional polymeric hydrogels and porous membranes. *J. Mater. Chem.* **2011**, *21*, 2783–2811. [CrossRef]
16. Salehi, S.M.; Di Profio, G.; Fontananova, E.; Nicoletta, F.P.; Curcio, E.; De Filpo, G. Membrane Distillation by Novel Hydrogel Composite Membranes. *J. Membr. Sci.* **2016**, *504*, 220–229. [CrossRef]
17. Di Profio, G.; Polino, M.; Nicoletta, F.P.; Belviso, B.D.; Caliandro, R.; Fontananova, E.; De Filpo, G.; Curcio, E.; Drioli, E. Tailored hydrogel membranes for efficient protein crystallization. *Adv. Funct. Mater.* **2014**, *24*, 1582–1590. [CrossRef]
18. Salehi, S.M.; Manjua, A.C.; Belviso, B.D.; Portugal, C.A.M.; Coelhoso, I.M.; Mirabelli, V.; Fontananova, E.; Caliandro, R.; Crespo, J.G.; Curcio, E.; et al. Hydrogel Composite Membranes Incorporating Iron Oxide Nanoparticles as Topographical Designers for Controlled Heteronucleation of Proteins. *Cryst. Growth Des.* **2018**, *18*, 3317–3327. [CrossRef]
19. Di Profio, G.; Salehi, S.M.; Caliandro, R.; Guccione, P.; Nico, G.; Curcio, E.; Fontananova, E. Bioinspired synthesis of CaCO$_3$ superstructures through a novel hydrogel composite membranes mineralization platform: A comprehensive view. *Adv. Mater.* **2016**, *28*, 610–616. [CrossRef] [PubMed]
20. Mirabelli, V.; Salehi, S.M.; Angiolillo, L.; Belviso, B.D.; Conte, A.; del Nobile, M.A.; Di Profio, G.; Caliandro, R. Enzyme Crystals and Hydrogel Composite Membranes as New Active Food Packaging Material. *Glob. Chall.* **2018**, *2*, 1700089. [CrossRef]
21. Israelachvili, J.; Pashley, R. The hydrophobic interaction is long range, decaying exponentially with distance. *Nature* **1982**, *300*, 341–342. [CrossRef]
22. Tsao, Y.; Evans, D.; Wennerstrom, H. Long-range attractive force between hydrophobic surfaces observed by atomic force microscopy. *Science* **1993**, *262*, 547–550. [CrossRef] [PubMed]
23. Meyer, E.E.; Rosenberg, K.J.; Israelachvili, J. Recent progress in understanding hydrophobic interactions. *Proc. Natl. Acad. Sci. USA* **2006**, *103*, 15739–15746. [CrossRef]
24. Lafuma, A.; Quere, D. Superhydrophobic states. *Nat. Mater.* **2003**, *2*, 457–460. [CrossRef]
25. Ma, Z.; Hong, Y.; Ma, L.; Su, M. Superhydrophobic membranes with ordered arrays of nanospiked microchannels for water desalination. *Langmuir* **2009**, *25*, 5446–5450. [CrossRef]
26. Zuo, G.; Wang, R. Novel membrane surface modification to enhance anti-oil fouling property for membrane distillation application. *J. Membr. Sci.* **2013**, *447*, 26–35. [CrossRef]
27. Wang, Z.; Elimelech, M.; Lin, S. Environmental applications of interfacial materials with special wettability. *Environ. Sci. Technol.* **2016**, *50*, 2132–3150. [CrossRef] [PubMed]
28. Wang, Z.; Jin, J.; Hou, D.; Lin, S. Tailoring surface charge and wetting property for robust oil-fouling mitigation in membrane distillation. *J. Membr. Sci.* **2016**, *516*, 113–122. [CrossRef]
29. Pashley, R.M. Hydration forces between mica surfaces in aqueous electrolyte solutions. *J. Colloid Interface Sci.* **1981**, *80*, 153–162. [CrossRef]
30. Chen, S.; Li, L.; Zhao, C.; Zheng, J. Surface hydration: Principles and applications toward low-fouling/nonfouling biomaterials. *Polymer* **2010**, *51*, 5283–5293. [CrossRef]

31. Tiraferri, A.; Kang, Y.; Giannelis, E.P.; Elimelech, M. Superhydrophilic thin-film composite forward osmosis membranes for organic fouling control: Fouling behavior and antifouling mechanisms. *Environ. Sci. Technol.* **2012**, *46*, 11135–11144. [CrossRef] [PubMed]

32. Fe, H.; Lu, D.; Cheng, W.; Zhang, T.; Lu, X.; Liu, Q.; Jiang, J.; Technology, P.; Cheng, W.; Zhang, T.; et al. Hydrophilic Fe_2O_3 dynamic membrane mitigating fouling of support ceramic membrane in ultrafiltration of oil/water emulsion. *Sep. Purif. Technol.* **2016**, *165*, 1–9.

33. Lin, S.; Nejati, S.; Boo, C.; Hu, Y.; Osuji, C.O.; Elimelech, M. Omniphobic membrane for robust membrane distillation. *Environ. Sci. Technol. Lett.* **2014**, *1*, 443–447. [CrossRef]

34. Boo, C.; Lee, J.; Elimelech, M. Engineering surface energy and nanostructure of microporous films for expanded membrane distillation applications. *Environ. Sci. Technol.* **2016**, *50*, 8112–8119. [CrossRef]

35. Wang, Z.; Hou, D.; Lin, S. Composite membrane with underwater-oleophobic surface for anti-oil-fouling membrane distillation. *Environ. Sci. Technol.* **2016**, *50*, 3866–3874. [CrossRef] [PubMed]

36. Lin, P.J.; Yang, M.C.; Li, Y.L. Prevention of surfactant wetting with agarose hydrogel layer for direct contact membrane distillation used in dyeing wastewater treatment. *J. Membr. Sci.* **2015**, *475*, 511–520. [CrossRef]

37. Ardeshiri, F.; Akbari, A.; Peyravi, M.; Jahanshahi, M. PDADMAC/PAA semi-IPN hydrogel-coated PVDF membrane for robust anti-wetting in membrane distillation. *J. Ind. Eng. Chem.* **2019**, *74*, 14–25. [CrossRef]

38. Talaty, E.R.; Raja, S.; Storhaug, V.J.; Dollle, A.; Carper, W.R. Raman and Infrared Spectra and ab Initio Calculations of C2–4MIM Imidazolium Hexafluorophosphate Ionic Liquids. *J. Phys. Chem. B* **2004**, *108*, 13177–13184. [CrossRef]

39. Paulechka, Y.U.; Kabo, G.J.; Blokhin, A.V.; Vydrov, O.A.; Magee, J.W.; Frenkel, M. Thermodynamic Properties of 1-Butyl-3-methylimidazolium Hexafluorophosphate in the Ideal Gas State. *J. Chem. Eng. Data* **2003**, *48*, 457–462. [CrossRef]

40. Peng, P.; Fane, A.G.; Li, X. Desalination by membrane distillation adopting a hydrophilic membrane. *Desalination* **2005**, *173*, 45–54. [CrossRef]

41. Hagmeyer, G.; Gimbel, R. Modelling the salt rejection of nanofiltration membranes for ternary ion mixtures and for single salts at different pH values. *Desalination* **1998**, *117*, 247–256. [CrossRef]

42. Garcia-Aleman, J.; Dickson, J.M. Permeation of mixed-salt solutions with commercial and pore-filled nanofiltration membranes: Membrane charge inversion phenomena. *J. Membr. Sci.* **2004**, *239*, 163–172. [CrossRef]

43. Schaep, J.; van der Bruggen, B.; Vandecasteele, C.; Wilms, D. Influence of ion size and charge in nanofiltration. *Sep. Purif. Technol.* **1998**, *14*, 155–162. [CrossRef]

44. Gan, Q.; Rooney, D.; Xue, M.; Thompson, G.; Zou, Y. An experimental study of gas transport and separation properties of ionic liquids supported on nanofiltration membranes. *J. Membr. Sci.* **2006**, *280*, 948–956. [CrossRef]

MDPI

St. Alban-Anlage 66

4052 Basel

Switzerland

Tel. +41 61 683 77 34

Fax +41 61 302 89 18

www.mdpi.com

ChemEngineering Editorial Office

E-mail: chemengineering@mdpi.com

www.mdpi.com/journal/chemengineering